SEVENTH EDITION

The American Class Structure in an Age of Growing Inequality

D1158894

SEVENTH EDITION

The American Class Structure in an Age of Growing Inequality

Dennis Gilbert

Hamilton College

PINE FORGE PRESS
An Imprint of Sage Publications, Inc.
Los Angeles • London • New Delhi • Singapore

For information:

Pine Forge Press
A Sage Publications Company
2455 Teller Road
Thousand Oaks,
California 91320
E-mail: order@sagepub.com

Sage Publications India Pvt. Ltd.
B 1/I 1 Mohan Cooperative
   Industrial Area
Mathura Road, New Delhi 110 044
India

Sage Publications Ltd.
1 Oliver's Yard
55 City Road
London EC1Y 1SP
United Kingdom

Sage Publications Asia-Pacific Pte. Ltd.
33 Pekin Street #02-01
Far East Square
Singapore 048763

Printed in the United States of America.

Library of Congress Cataloging-in-Publication Data

Gilbert, Dennis L.
The American class structure in an age of growing inequality / Dennis Gilbert. — 7th ed.
  p. cm.
Includes bibliographical references and index.
ISBN 978-1-4129-5414-3 (pbk.)
  1. Social classes—United States. 2. Equality—United States. 3. Poverty—United States. 4. United States—Social conditions. 5. United States—Economic conditions. I. Title.

HN90.S6G54 2008
305.50973—dc22   2007039090

This book is printed on acid-free paper.

08   09   10   11   12   10   9   8   7   6   5   4   3   2   1

Acquisitions Editor:	Benjamin Penner
Editorial Assistant:	Nancy Scrofano
Production Editor:	Tracy Buyan
Copy Editor:	Teresa Herlinger
Typesetter:	C&M Digitals (P) Ltd.
Proofreader:	Wendy Jo Dymond
Indexer:	Maria Sosnowski
Cover Designer:	Gail Buschman
Marketing Manager:	Jennifer Banando

For Eric, Christi, Trevor, Allison, and David.
August and October 2007

May their memory be a comfort to those who loved them.

Contents

About the Author

Dennis Gilbert is Professor of Sociology at Hamilton College. He holds a PhD from Cornell University and has been a visiting professor at Cornell and at the Universidad Católica in Lima, Peru. He is also the author of *Mexico's Middle Class in the Neoliberal Era, Sandinistas: The Party and the Revolution,* and *La Oligarquia Peruana: Historia de Tres Familias.*

Preface

I was 12 years old when the original version of *The American Class Structure* was being written in 1955. The author was Joseph Kahl, an unemployed Harvard PhD then living cheaply in Mexico. His book, which helped define the emerging field of social stratification, remained in print, without revision, for 25 years. It earned this long run by presenting a lucid synthesis of the best research on the American class system. Each study was lovingly dissected by Kahl, who conveyed its flavor, assessed its strengths and weaknesses, summarized its most significant conclusions, and explained how they were reached.

The American Class Structure was not a theoretical book. Kahl created a simple conceptual schema with a short list of key variables drawn from the work of Karl Marx and Max Weber. Kahl admitted that he had settled on this framework for the good and practical reason that it allowed him to draw together the results of disparate research reports. But the variables were interrelated, and Kahl believed that they tended to converge to create social classes in a pattern he called the American class structure. At the same time, he recognized that classes and class structure were abstractions from social reality—tendencies never fully realized in any situation but discernable when one stepped back from detail to think about underlying forces.

Sometime around 1980, Kahl invited me to collaborate on a new version of *The American Class Structure*. He was then professor of sociology at Cornell, and I was his graduate student. The book we published in 1982 encompassed a body of stratification research that had grown enormously in sophistication and volume since the 1950s. *The American Class Structure: A New Synthesis* consisted almost entirely of new material but preserved the general framework of the original edition and its analyses of classic studies of the American class system. That edition and two subsequent editions, which Kahl and I produced together, proved popular with a new generation of sociologists and sociology students.

But when our publisher asked us to start thinking about another edition, Kahl, who had by then retired to Chapel Hill, North Carolina, indicated that he would rather spend his time listening to opera than reading page proofs again. And since he would not be contributing to the new edition, he asked that his name be taken off the cover. Thus, the subsequent editions have been published under my name.

Although there is now only one official author, the authorial "I" reverts to "we" after this preface. Much of this book is the product of a long collaboration, and I am often at a loss to recall who wrote (or perhaps rewrote) a particular passage. Retaining the "we" of earlier editions seemed perfectly natural. That said, I want to stress that I bear sole responsibility for every word included in this edition.

I am, in particular, responsible for the theme that runs through the recent editions and is reflected in the revised subtitle: *In an Age of Growing Inequality.* This theme was inspired by data on trends in earnings, income, wealth, and related variables that reveal a remarkably consistent pattern of rising class inequalities since the mid-1970s. This pattern sharply contrasts with the broadly shared prosperity of the 1950s and 1960s. The text repeatedly returns to a deceptively simple question: Why is this happening?

Like its predecessors, the seventh edition of *The American Class Structure* is not an encyclopedic survey of stratification research, nor is it an exercise in class theory. It revolves around a short list of variables, largely derived from classical theory; emphasizes selected empirical studies; and focuses on the socioeconomic core of the class system. Gender and race are treated in relation to class, rather than as parallel dimensions of stratification. This approach reveals that the experience of class is inextricably bound up with gender and race. For example, studies show that a married woman's sense of class identity reflects her husband's job, her own job, and her attitude toward gender roles. Residential segregation by class is increasing in the United States, but so is segregation by race. One result has been the growth of affluent black neighborhoods.

For this edition, I have made revisions to every chapter, adding new material on income, wealth, earnings, jobs, poverty, gender, race, politics, and other topics. I have especially focused on information that might tell whether the late–twentieth century trend toward growing inequality is continuing into the twenty-first century. This edition includes important new work on class patterns in child rearing, fresh estimates of social mobility (it's declining), and a 10-year assessment of the 1996 welfare reform. As I have added new material, I have eliminated or summarized our coverage of older studies.

Two well-received features of the last edition have been retained. One is streamlined citation of statistical sources, so as to produce a less cluttered text. On this feature, see the "Note on Statistical Sources" at the end of the book, which describes the principal sources of statistics on income, occupation, and related topics. The other is the glossary, added to make life easier for readers who are puzzled by Marx's use of the term *ideology,* who are uncertain about the exact meaning of net worth, or who can't recall how the text defined postindustrial society. The glossary has been revised, where appropriate, for this edition. Readers will again find a list of relevant glossary terms at the end of each chapter.

In earlier prefaces, Kahl and I thanked many friends, colleagues, and students whose help made *The American Class Structure* a better book. Here I want to acknowledge Scott Sernau's help with the glossary and the research and editing assistance provided by Kate Speirs, Grace Dobbyn, and Robin Vanderwall. I am grateful to Ben Penner of Pine Forge Press for his enthusiastic support of this new edition and to the Pine Forge reviewers listed below for their valuable suggestions. Of course, I am, as ever, thankful to Joe Kahl, a fine teacher, supportive colleague, and good friend.

Dennis Gilbert
Clinton, New York

Pine Forge Reviewers

Scott Appelrouth
California State University,
Northridge

Jane Emery Prather
California State University,
Northridge

E. C. Ejiogu
University of Maryland,
College Park

Elizabeth Stearns
University of North Carolina
at Charlotte

Kenneth J. Herrmann, Jr.
SUNY Brockport

Mike Weissbuch
Xavier University

David R. Maines
Oakland University

Thomas C. Wilson
Florida Atlantic University

Roxana Moayedi
Trinity University

Anne Wortham
Illinois State University

Joan Olson
University of Mary Washington

CHAPTER 1

Social Class in America

All communities divide themselves into the few and the many. The first are the rich and well-borne, the other the mass of the people. . . . The people are turbulent and changing; they seldom judge or determine right. . . . Give, therefore, to the first class a distinct, permanent share in the government. They will check the unsteadiness of the second, and as they cannot receive any advantage by a change, they therefore will ever maintain good government.

Alexander Hamilton (1780)

On the night the *Titanic* sank on her maiden voyage across the Atlantic in 1912, social class proved to be a key determinant of who survived and who perished. Among the women (who were given priority over men for places in the lifeboats), 3% of the first-class passengers drowned, compared with 16% of the second-class and 45% of the third-class passengers. Of the victims in first class, all but one had refused to abandon ship when given the opportunity. On the other hand, third-class passengers had been ordered to stay below deck, some of them at the point of a gun (Lord 1955:107, cited in Hollingshead and Redlich 1958:6).

The divergent fates of the *Titanic*'s passengers present a dramatic illustration of the connection between social class and what pioneer sociologist Max Weber called *life chances.* Weber invented the term to emphasize the extent to which our chances for the good things in life are shaped by class position.

Contemporary sociology has followed Weber's lead and found that the influence of social class on our lives is indeed pervasive. Table 1.1 gives a few examples. These statistics compare people at the bottom, middle, and top of the class structure. They show, among other things, that people in the bottom 25% are less likely to be in good health, more likely to find life boring, less likely to have Internet access, and more likely to be the victims of violent crime. Those in the top 25% are healthier, safer, more likely to send their kids to college, and more optimistic about the future. It is no wonder that they are, on average, happier with their lives.

Thoughtful observers have recognized the importance of social classes since the beginnings of Western philosophy. They knew that some individuals and families had more money, more influence, or more prestige than their neighbors. The philosophers also realized that the differences were more than personal or even familial, for the pattern of inequalities tended to congeal into strata of families who shared similar positions. These social strata or classes divided society into a hierarchy;

Table 1.1   Life Chances by Social Class[a]

	Bottom	Middle	Top
General health excellent[b]	56%	82%	91%
Victims of violent crime/1000 pop.[c]	44	34	26
Own home[b]	35%	64%	88%
Children 3–17 with home internet access[d]	34%	70%	93%
Children 18–24 in college or college grad.[e]	30%	52%	72%
Find life "exciting" (not "routine" or "dull")[b]	36%	45%	58%
See opportunities to get ahead[b]	66%	79%	85%
Very happy[b]	20%	31%	43%

a. Classes defined by income: Bottom 25%, middle 50%, and top 25%.
b. General Social Survey 2000. Computed for this table.
c. U.S. Department of Justice, Bureau of Justice Statistics 2001.
d. In 2003. Calculated from U.S. Census statistics.
e. In 2005. Calculated from U.S. Census statistics.

each stratum had interests or goals in common with equals but different from, and often conflicting with, those of groups above or below them. Finally, it was noted that political action often flows from class interests. As one of the founding fathers, Alexander Hamilton observed, the rich seek social stability to preserve their advantages, but the poor work for social change that would bring them a larger share of the world's rewards.

This book is an analysis of the American class system. We explore class differences in income, prestige, power, and other key variables. We will point out how these variables react on one another—for instance, how a person's income affects beliefs about social policy or how one's job affects the choice of friends or spouse. And we will explore the question of movement from one class to another, recognizing that a society can have classes and still permit individuals to rise or fall among them.

We begin by consulting two major theorists of social stratification, Karl Marx and Max Weber, to identify the major facets of the subject as a guide. Marx (1818–1883) and Weber (1864–1920) established an intellectual framework that strongly influenced subsequent scholars. (Social stratification, by the way, refers to social ranking based on characteristics such as wealth, occupation, or prestige.)

Karl Marx

Although the discussion of stratification goes back to ancient philosophy, modern attempts to formulate a systematic theory of class differences began with Marx's work in the nineteenth century. Most subsequent theorizing has represented an attempt either to reformulate or to refute his ideas. Marx, who was born in the wake of the French Revolution and lived in the midst of the Industrial Revolution, emphasized the study of social class as the key to an understanding of the turbulent events of his time. His studies of economics, history, and philosophy convinced him that societies are mainly shaped by their economic organization and that social classes form the link between economic facts and social facts. He also concluded that fundamental social change is the product of conflict between classes. Thus, in Marx's view, an understanding of classes is basic to comprehending how societies function and how they are transformed.

In Marx's work, social classes are defined by their distinctive relationships to the means of production. Taking this approach, Marx defined two classes in the emerging industrial societies of his own time: the capitalist class (or *bourgeoisie*) and the working class (or *proletariat*). He describes the bourgeoisie as the class that owns the means of production, such as mines or factories, and the proletariat as the class of those who must sell their labor to the owners of the means to earn a wage and stay alive. Marx maintained that in modern, capitalist society, each of these two basic classes tends toward an internal homogeneity that obliterates differences within them. Little businesses lose out in competition with big businesses, concentrating ownership in a small bourgeoisie of monopoly capitalists. In a parallel fashion, machines get more sophisticated and do the work that used to be done by skilled workers, so gradations within the proletariat fade in significance.

As the basic classes become internally homogenized, the middle of the class structure thins out and the system as a whole becomes polarized between the two class extremes.

But notice that these broad generalizations refer to long-range trends. Marx recognized that at any given historical moment, the reality of the class system was more complex. The simplifying processes of homogenization and polarization were tendencies, unfolding over many decades, which might never be fully realized. Marx's descriptions of contemporary situations in his writings as a journalist and pamphleteer show more complexity in economic and political groupings than do his writings as a theorist of long-term historical development.

We have noted that Marx defined the proletariat, bourgeoisie, and other classes by their relationship to the means of production. Why? In the most general sense, because he regarded production as the center of social life. He reasoned that people must produce to survive, and they must cooperate to produce. The individual's place in society, relationships to others, and outlook on life are shaped by his or her work experience. More specifically, those who occupy a similar role in production are likely to share economic and political interests that bring them into conflict with other participants in production. Capitalists, for instance, reap profit (in Marx's terms, *expropriate surplus*) by paying their workers less than the value of what they produce. Therefore, capitalists share an interest in holding down wages and resisting legislation that would enhance the power of unions to press their demands on employers.

From a Marxist perspective, the manner in which production takes place (that is, the application of technology to nature) and the class and property relationships that develop in the course of production are the most fundamental aspects of any society. Together, they constitute what Marx called the *mode of production.* Societies with similar modes of production ought to be similar in other significant respects and should therefore be studied together. Marx's analysis of European history after the fall of Rome distinguished three modes of production, which he saw as successive stages of societal development: *feudalism,* the locally based agrarian society of the Middle Ages, in which a small landowning aristocracy in each district exploited the labor of a peasant majority; *capitalism,* the emerging industrial and commercial order of Marx's own lifetime, already international in scope and characterized by the dominance of the owners of industry over the mass of industrial workers; and *communism,* the technologically advanced, classless society of the future, in which all productive property would be held in common.

Marx regarded the mode of production as the main determinant of a society's *superstructure* of social and political institutions and ideas. He used the concept of superstructure to answer an old question: How do privileged minorities maintain their positions and contain the potential resistance of exploited majorities? His reply was that the class that controls the means of production typically controls the means of compulsion and persuasion—the superstructure. He observed that in feudal times, the landowners monopolized military and political power. With the rise of modern capitalism, the bourgeoisie gained control of political institutions. In each case, the privileged class could use the power of the state to protect its own interests. For instance, in Marx's own time, the judicial, legislative, and police authority of European governments dominated by the bourgeoisie were employed to crush the early labor movement, a pattern that was repeated a little later in the United States. In an insightful overstatement from the *Communist Manifesto* (1848), Marx asserted, "The executive of the modern State is but a committee for managing the common affairs of the whole bourgeoisie" (Marx 1979:475).

But Marx did not believe that class systems rested on pure compulsion. He allowed for the persuasive influence of ideas. Here Marx made one of his most significant contributions to social science: the concept of *ideology*. He used the term to describe the pervasive ideas that uphold the *status quo* and sustain the ruling class. Marx noted that human consciousness is a social product. It develops through our experience of cooperating with others to produce and to sustain social life. But social experience is not homogeneous, especially in a society that is divided into classes. The peasant does not have the same experience as the landlord and therefore develops a distinct outlook. One important feature of this differentiation of class outlooks is the tendency for members of each group to regard their own particular class interests as the true interests of the whole society. What makes this significant is that one class has superior capacity to impose its self-serving ideas on other classes.

The class that dominates production, Marx argued, also controls the institutions that produce and disseminate ideas, such as schools, mass media, churches, and courts. As a result, the viewpoint of the dominant class pervades thinking in areas as diverse as the laws of family life and property, theories of political democracy, notions of economic rationality, and even conceptions of the afterlife. In Marx's (1979) words, "the ideas of the ruling class are in every epoch the ruling ideas" (p. 172). In extreme situations, ideology can convince slaves that they ought to be obedient to their masters, or poor workers that their true reward will eventually come to them in heaven.

Marx (1979) maintained, then, that the ruling class had powerful political and ideological means to support the established order. Nonetheless, he regarded class societies as intrinsically unstable. In a famous passage from the *Communist Manifesto*, he observed,

> The history of all hitherto existing society is the history of class struggles. Freeman and slave, patrician and plebeian, lord and serf, guildmaster and journeyman, in a word, oppressor and oppressed stood in constant opposition to one another, carried on an uninterrupted, now hidden, now open fight, a fight that each time ended either in a revolutionary reconstitution of society at large, or in the common ruin of the contending classes.
>
> In the earlier epochs of history, we find almost everywhere a complicated arrangement of society into various orders, a manifold gradation of social rank. In ancient Rome, we have patricians, knights, plebeians, slaves; in the Middle Ages, feudal lords, vassals, guild-masters, journeymen, apprentices, serfs; in almost all of these classes, again, subordinate gradations. . . .
>
> Our epoch, the epoch of the bourgeoisie, possesses, however, this distinctive feature: It has simplified the class antagonisms. Society as a whole is more and more splitting up into two great hostile camps, into two great classes directly facing each other: Bourgeoisie and Proletariat. (Pp. 473–474)

As these lines suggest, Marx saw class struggle as the basic source of social change. He coupled class conflict to economic change, arguing that the development of new means of production (for example, the development of modern industry) implied the emergence of new classes and class relationships. The most serious political conflicts develop when the interests of a rising class are opposed to those of an established ruling class. Class struggles of this sort can produce a "revolutionary reconstitution of society." Notice that each epoch creates within itself the growth of a new class that eventually seizes power and inaugurates a new epoch.

Two eras of transformation through class conflict held particular fascination for Marx. One was the transition from feudalism to modern capitalism in Europe, a process in which he assigned the bourgeoisie (the urban capitalist class) "a most revolutionary part" (Marx 1979:475). Into a previously stable agrarian society, the bourgeoisie introduced a stream of technological innovations, an accelerating expansion of production and trade, and radically new forms of labor relations. The feudal landlords, feeling their own interests threatened, resisted change. The result was a series of political conflicts (the French Revolution was the most dramatic instance) through which the European bourgeoisie wrested political power from the landed aristocracy.

Marx believed that a second, analogous era of transformation was beginning during his own lifetime. The capitalist mode of production had created a new social class, the urban working class, or proletariat, with interests directly opposed to those of the dominant class, the bourgeoisie. This conflict of interests arose, not simply from the struggle over wages between capital and labor, but from the essential character of capitalist production and society. The capitalist economy was inherently unstable and subject to periodic depressions with massive unemployment. These economic crises heightened awareness of long-term trends widening the gap between rich and poor. Furthermore, capitalism's blind dependence on market mechanisms built on individual greed created an alienated existence for most members of society. Marx was convinced that only under communism, with the means of production communally controlled, could these conditions be overcome.

The situation of the proletarian majority made it capitalism's most deprived and alienated victim and therefore the potential spearhead of a communist revolution. However, in Marx's view, an objective situation of class oppression does not lead directly to political revolt. For that to happen, the oppressed class must first develop *class consciousness*—that is, a sense of shared identity and common grievances, requiring a collective response. Some of Marx's most fruitful sociological work, to which we will return in Chapter 9, is devoted to precisely this problem. What intrinsic tendencies of capitalist society, Marx asked, are most likely to produce a class-conscious proletariat? Among the factors he isolated were the stark simplification of the class order in the course of capitalist development; the concentration of large masses of workers in the new industrial towns; the deprivations of working-class people, exacerbated by the inherent instability of the capitalist economy; and the political sophistication gained by the proletariat through participation in working-class organizations such as labor unions and mass political parties.

What, in sum, can be said of Marx's contribution to stratification theory? His recognition of the economic basis of class systems was a crucial insight. His theory of ideology and his conception of the connection between social classes and political processes, although oversimple as stated, proved a fruitful starting point for modern research. As for his conception of change, a series of twentieth-century revolutions—including those in Mexico (1910), Russia (1917), and China (1949)—have established the significance of class conflict for radical social transformation. However, social revolutions have typically occurred in peasant societies during early stages of industrialization under foreign influence rather than in the advanced industrial countries where Marx anticipated them. In the advanced industrial countries, the proletariat used labor unions and mass political parties to defend its interests, thus rechanneling the forces of class conflict into the legal procedures of democratic politics.

A century after his death, it is apparent that Marx was a better sociologist than he was a prophet. He identified many of the central processes of capitalist society, but he was unable to foresee all the consequences of their unfolding, and his vision of a humane socialist future has not been realized in any communist country.

Max Weber

The great German sociologist Max Weber, who wrote in the early years of the twentieth century, was interested in many of the same problems that had fascinated Marx—among them, the origins of capitalism, the role of ideology, and the relationship between social structure and economic processes. Weber frequently benefited from Marx's work, even while reaching rather different conclusions. In the field of stratification, his special contributions were (1) to introduce a conceptual clarity that was often lacking in Marx's references to social classes and (2) to highlight the subjective aspects of stratification, as expressed in everyday interactions.

Weber made a crucial distinction between two orders of ranking or stratification: *class* and *status*. Class had roughly the same meaning for both Weber and Marx. It refers to groupings of people according to their economic position. Class situation or membership, according to Weber, is defined by the individual's strength in economic markets (for example, the job market or securities markets), to the extent that these determine individual *life chances*. By life chances, he meant the fundamental aspects of an individual's future possibilities that are shaped by class membership, from the infant's chances for decent nutrition to the adult's opportunities for worldly success.

Following Marx, Weber stressed that the most important class distinction is between those who own property and those who do not. However, he noted that many significant distinctions could be made within these two categories. Among the propertied elite, for example, there are rentiers, who support themselves with income from stocks, bonds, and other securities, and entrepreneurs, who depend on profits from businesses they own and operate. The propertyless can be differentiated by the occupational skills that they bring to the marketplace: The life chances of an unskilled worker are vastly different from those of a well-trained engineer. This suggests that the vast population of wage earners whom Marx lumped into the proletariat were really a highly differentiated group.

A *social class,* then, becomes a group of people who share the same economically shaped life chances. Notice that this way of defining a class does not imply that the individuals in it are necessarily aware of their common situation. It simply establishes a statistical category of people who are, from the point of view of the market (and the sociologist), similar to each other. Only under certain circumstances do they become aware of their common fate, begin to think of each other as equals, and develop institutions of joint action to further their shared interests.

Status, the second major order of stratification defined by Weber, is ranking by social prestige. In contrast with class, which is based on objective economic fact, status is a subjective phenomenon, a sentiment in people's minds. Although the members of a class may have little sense of shared identity, the members of a status group generally think of themselves as a social community, with a common *lifestyle* (a familiar term we owe to Weber). In a classic essay on stratification, Weber (1946) outlined these distinctions:

> In contrast to the purely economically determined "class situation," we wish to designate as "status situation" every typical component of the life fate of men that is determined by a specific, positive or negative, social estimation of honor. . . .
>
> Status groups are normally communities. They are, however, often of an amorphous kind. . . . In content, status honor is normally expressed by the fact that above all else a specific style of life can be expected from all those who wish to belong to the circle. Linked with this expectation are restrictions on "social" intercourse (that is, intercourse which is not subservient to economic or any other of business's "functional" purposes). These restrictions may confine normal marriages to within the status circle and may lead to complete endogamous closure. . . .
>
> Of course, material monopolies provide the most effective motives for the exclusiveness of a status group. . . . With an increased inclosure of the status group, the conventional preferential opportunities for special employment grow into a legal monopoly of special offices for the members. . . .
>
> With some over-simplification, one might thus say that "classes" are stratified according to their relations to the production and acquisition of goods; whereas "status groups" are stratified according to the principles of their consumption of goods as represented by special "styles of life." (Pp. 186–193)

In those passages, Weber specified many of the interrelations between class and status, between economy and society. Because of class position, a person earns a certain income. That income permits a certain lifestyle, and people soon make friends with others who live the same way. As they interact with one another, they begin to conceive of themselves as a special type of people. They restrict interaction with outsiders who seem too different (they may be too poor, too uneducated, too clumsy to live graciously enough for acceptance as worthy companions). Marriage partners are chosen from similar groups because once people follow a certain style of life, they find it difficult to be comfortable with people who live differently. Thus, the status group becomes an ingrown circle. It earns a position in the local community that entitles its members to social honor or prestige from inferiors.

Status groups develop the conventions or customs of a community. Through time, they evolve appropriate ways of dressing, of eating, and of living that are somewhat different from the ways of other groups. These ways are expressed as moral judgments reflecting abstract principles of value that separate "good" from "bad." The application of these principles to individuals establishes rankings of social honor or prestige. These distinctions often react back on the marketplace; to preserve their advantages, high-status groups attempt to monopolize those goods that symbolize their style of life—they pass consumption laws prohibiting the lower orders from wearing lace, or they band together to keep Jews or blacks out of prestigious country clubs. (Weber regarded invidious distinctions among ethnic groups as a type of status stratification.)

A status order tends to restrict the freedom of the market, not only by its monopolization of certain types of consumption goods, but also by its monopolization of the opportunities to earn money. If they can get the power, status groups often restrict entry into the more lucrative professions or trades and access to credit. For example, entry into the electricians' union might be restricted to sons of current members. The local bank might be more willing to grant a loan to a member of the country club than to a social nobody. More generally, birth into a high-status family gives children advantages of social grace and personal contacts that eventually help their careers.

Weber observed that, in theory, class and status are opposed principles. In its purest form, the class or economic order is universalistic and impersonal; it recognizes no social distinctions and judges solely on the basis of competitive skill or accumulated wealth. Status, in contrast, is based on particularistic distinctions: some people are "better" than others. Status groups want to restrict freedom of competition in both production and consumption.

Weber recognized that, in practice, class and status are intertwined—at least in the long run. Historically, the status order is created by the class order; consumption, after all, is based on production. For example, although elite society might react against the status claims of the newly rich, it typically accepts their descendants if they have properly cultivated the conventions of the higher status group. On another level, the appearance of classes based on new sources of wealth—for instance, the emergence of an industrial bourgeoisie in Europe and America in the nineteenth century—signals a future restructuring of the status order as a whole.

Weber, like Marx, was interested in the relationship between stratification and political power. It would be accurate to say that for both men, stratification was essentially a political topic. But Weber was highly skeptical of the implication in Marx's work that all political phenomena could be traced back directly to class. For instance, Weber suggested that the institutions of the modern bureaucratic state exercise an influence on society that is not reducible to the control exercised by a single class. (In *The Eighteenth Brumaire,* Marx [1979:594–617] himself reluctantly adopted this view for a special circumstance, but not Weber's corollary that a communist state might be grimly similar to a capitalist state, reflecting bureaucratic domination of society.)

Weber opposed what he called the "pseudo-scientific operation" of Marxist writers of his day, who assumed an automatic link between class position and class consciousness (Weber 1946:184). He noted that a shared economic situation can and sometimes does lead to an awareness of shared class interests and a willingness to engage in militant class action, but it need not. Indeed, the very notion of class interest was highly ambiguous for Weber. In his view, there are multiple classes in modern societies and they are continually changing. Under such conditions, individuals may think of their own identities, and shared or conflicting interests with others, in varied ways. Someone whom sociologists would identify as working class might think of himself as white and middle class, because he believes he has nothing in common with minority workers and supposes himself to be a middle-income, average American. Or, he might strongly identify with other workers, whatever their race, and become class conscious in the Marxian sense. Neither would surprise Weber.

Implicit in Weber's approach to stratification is the idea that status considerations can undermine the development of class consciousness and class struggle. For example, the politics of the American South has long been shaped by the tendency of poor whites to identify with richer whites rather than with poor blacks who share their economic position. Weber noted that political parties can develop around class, status, or other bases for conflict over power. The major American political parties are amorphous coalitions that have never been as clearly oriented toward the pursuit of class interests as have, for example, the working-class parties of Western Europe. None of this would have surprised Weber.

In sum, Weber accepted Marx's idea of the underlying economic basis of stratification. But Weber's conception of social class was much more flexible than Marx's and probably better adapted to the complexities of modern societies. Weber also identified another order of stratification, by differentiating between class and status. He argued that the two interact with each other and with the political process in ways not fully recognized by Marx.

Three Issues and Ten Variables

Marx's and Weber's writings suggest three broad issues in the study of social class:

1. *Economic basis.* How do class distinctions arise from economic distinctions? And how, in particular, does economic change transform the class system? These were central concerns for both theorists.

2. *Social basis.* How are economic class distinctions reflected in social distinctions and social behavior? Weber's discussion of status groups is relevant here. He noted their tendency to become social communities with distinctive lifestyles and values. He was intrigued by the complex relationship between class and status.

3. *Political implications.* How does the class system affect the political system? How do economically dominant classes interact politically with the other classes in a society? For both Marx and Weber, class was ultimately a political topic.

These issues led us to organize our examination of the class system around a series of related sociological variables. With regard to the economic issue, we will be looking at *occupation, wealth, income,* and *poverty;* with regard to the social issue, at *prestige, association, socialization,* and *social mobility;* and with regard to the political issue, at *power* and *class consciousness.* (Association refers to the patterned social connections among people and socialization to the process though which the young learn the skills they need to participate in society. These and the other variables are more precisely defined in the appropriate chapters and in the glossary at the end of the book).

What Are Social Classes?

We define social classes as groups of families more or less equal in rank and differentiated from other families above or below them with regard to characteristics such as occupation, income, wealth, and prestige.[1]

Our approach raises two questions: Why conceive of stratification in terms of discrete *classes?* And why think of classes as groupings of *families?* The first question arises because it is logically possible for a society to be stratified in a continuous gradation between high and low without any sharp lines of division. In reality, this is unlikely. The sources of a family's position are shared by many other similar families; there is only a limited number of types of occupations or of possible positions in the property system. One holds a routine position in a service, factory, or office setting; lives by manual skill or professional expertise; or manages people and money. People in similar positions have similar incomes and a tendency to mix with one another, to grow similar in their thinking and lifestyle. The similarities are shared within families and often inherited by children. In other words, the various stratification variables tend to converge and jell; they form a pattern within which social classes begin to form.

The pattern formed by the objective connections among the variables is heightened by the way people think about social matters because popular thought tends toward stereotypes. Doctors are viewed as a homogeneous group, and distinctions among them tend to be ignored. Similarly, poor people tend to lump together all bosses, and rich people overlook the many distinctions that exist among those who labor for an hourly wage.

The second question arises because it is logically possible to study stratification of individuals rather than families. Why not define classes as discrete groups of *individuals* of equal rank? The simple answer, implied earlier, is that the members of a household live under the same roof, pool their resources, share a common economic fate, and tend, for all these reasons, to have a similar perspective on the world. This answer is not quite as persuasive as it was 30 or 40 years ago. What made it seem self-evident in the past was that families were largely dependent on income produced by a "male head of household." The sociologist could place the family in the class system on the basis of his occupation, which tended to be a good predictor of the family's

---

[1]At times we will use the terms *family* and *household* interchangeably. When discussing income, we will often make the Census Bureau's more rigorous distinction between a group of related people residing together (family) and the broader concept that encompasses families, individuals residing alone, and unrelated individuals residing together (household).

economic condition and its political outlook. Women, of course, were largely ignored in this conception of the class order, but it was arguably a reasonable approach to a world in which women's public economic role was quite circumscribed.

In the last few decades, women's economic and family roles have changed radically. Single women head a growing proportion of households. Married-couple families increasingly depend on two incomes. Where do we place a family in the class hierarchy if two spouses, both employed in working-class jobs, together produce a comfortable middle-class income? Suppose the husband is a factory worker and the wife is a teacher. Again, where do we place them? And, if occupation is the key to political outlook, whose occupation counts here?

There are no fully satisfactory answers to such questions—the world is a complicated place. But there are, again, tendencies toward convergence and consistency. Family members (whatever disparities exist among them) still depend on common resources. They are viewed by outsiders as sharing the same position within the community. Husbands and wives typically have similar levels of education and, as a result, there is a correlation between the jobs held by working couples. Although married-couple families have grown increasingly dependent on wives' earnings, husbands are still the most important providers in the great majority of families. In the chapters that follow, we will repeatedly return to these issues, exploring the changing economic role of women in some detail. But we will continue to regard families as the basic unit of stratification analysis and define classes as groups of families or households.

At the same time, we will use the term "family" in the broadest possible sense to include households consisting of one person and larger domestic units "headed" by single females, single males, or couples (both heterosexual and homosexual). We will generally establish the class position of a family by the occupation of the family member who is the principal income earner. (In some cases, we will use household wealth or dependence on government payments to define class position.)

In sum, we will interpret the stratification system with the 10 variables mentioned earlier and discrete social classes composed of families. But we recognize that households may have inconsistent scores on the variables (for example, a high-income, low-prestige occupation), that the lines dividing classes may be inconveniently fuzzy, and that the class placement of some families may be ambiguous. The reason for all this incoherence is not so much the inadequacy of the variables or definitions we use as the vague, fluid character of the stratification system itself. This book emphasizes the tendencies toward convergence, toward crystallization of the pattern, despite the many disturbing influences, often the result of social change, that keep the patterns from becoming as clear-cut in reality as in theory.

An American Class Structure

Some readers will have concluded by now that there is as much art as science in the study of social stratification—and they are probably right. We can make factual statements about, say, the distribution of income or patterns of association. But efforts to combine such information into broader statements about the class system run up against the inherent inconsistencies of social reality and are inevitably influenced by the viewpoint of the author.

We will, for example, be examining several general models of the class structure. Each tells us how many classes there are, how they can be distinguished from one another, and who belongs in each class. Some class models are more convincing than others because they make better use of the facts and illuminate matters that concern us. Some are obviously worthless. But there is really no way to distinguish the one "true" model.

Our own model of the American class structure represents a synthesis of what we have learned writing this book. We summarize it here and reconsider it in greater detail in the last chapter. The model, diagrammed in Figure 1.1, stratifies the population into six classes. The diagram shows the occupations and incomes typical of each class. The occupation referred to is that of a household's principal earner. The income is total household income, though, as will be seen, we are more interested in the principal sources of income rather than the level of income.

Drawing from Marx, we distinguish a very small top class, whose income derives largely from return on assets—the *capitalist class*. These are people who own lucrative businesses, commercial real estate, and securities such as stocks and bonds. They may hold jobs—some are top corporate executives—but ownership is the key to their

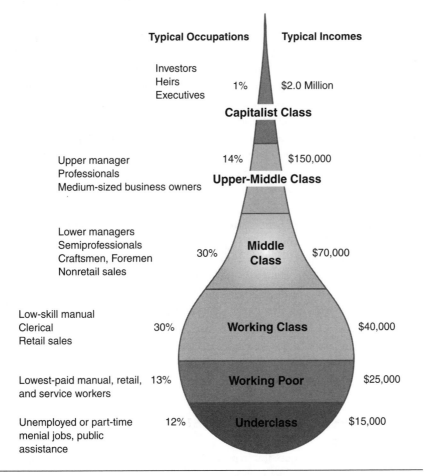

Typical Occupations   **Typical Incomes**

Investors
Heirs   1% $2.0 Million
Executives

Capitalist Class

Upper manager   14% $150,000
Professionals
Medium-sized business owners   **Upper-Middle Class**

Lower managers
Semiprofessionals   30%   **Middle** $70,000
Craftsmen, Foremen   **Class**
Nonretail sales

Low-skill manual
Clerical   30%   **Working Class** $40,000
Retail sales

Lowest-paid manual, retail,   13%   **Working Poor** $25,000
and service workers

Unemployed or part-time   12%   **Underclass** $15,000
menial jobs, public
assistance

Figure 1.1   Gilbert-Kahl Model of the Class Structure

high incomes. Drawing from Weber, we recognize multiple class distinctions among the largely nonpropertied majority—those who cannot live off what they own.

Below the capitalist class is an *upper-middle class* of well-paid, university-educated managers and professionals: people with responsible positions in business organizations, along with lawyers, doctors, accountants, and other specialists.

Next are the two largest classes, the *middle class* and the *working class*. Among those we place in the middle class are lower-level managers, insurance agents, teachers, nurses, electricians, and plumbers. Our working class includes unskilled factory workers, office workers without specialized training, and many retail sales workers. The boundary between these two classes cannot be sharply drawn. Note that we do not depend on the traditional blue collar–white collar (manual versus nonmanual) distinction; there are blue-collar and white-collar workers in both classes. Instead, we have delineated the two classes based on the levels of skill or knowledge and independence or authority associated with occupations.

At the bottom of the class structure are the *working poor* and the *underclass*. The working poor are employed at very low-skill, low-wage, often insecure jobs that do not pay benefits such as medical insurance. Fast-food workers, maids and janitors, and many unskilled construction workers fall into this class. Because their jobs are poorly paid, precarious, and devoid of benefits, their lives are marked by financial instability. Members of the underclass may have some job income, but they are often dependent on income from government programs, including social security, public assistance, and veterans benefits. A few draw income from criminal activities.

Note that this model or map of the class system is based entirely on economic distinctions. We do not incorporate prestige differences (in Weber's terms, status distinctions) because we believe they derive, in the long run, from economic differences. The model is built around *sources of income:* The top class draws income from capitalist property, the intermediate classes rely on earnings from jobs at differing occupational levels, and the bottom class depends on a mix of unstable job income and government payments. The emphasis here is on the *source* rather than the *level* of income. In fact, there is bound to be some overlap in income level between classes as defined here. In the middle of the model, *occupation* is the decisive variable, separating those who depend on jobs into distinct levels.

A final observation: The distinction between middle class and working class—traditionally portrayed by division between office and factory—was long regarded as the critical dividing line in the class structure. But today many office jobs are simplified and routinized like jobs in the factory. We believe that the line dividing the capitalist and upper-middle classes from the classes below them has become the most important boundary. One reason is that economic returns on capitalist property and on the advanced education typical of the upper-middle class have grown rapidly in recent years, while rewards for lower levels of education or skill have stagnated or shrunk.

Is the American Class Structure Changing?

We will return to this question repeatedly as we move from topic to topic. In particular, we will want to find out how the transformation of the U.S. economy in the last two or three decades has affected the class structure. In recent years, increasing class inequality has become a national political issue. Critics argue that the United States is becoming a less egalitarian, more rigidly stratified society. They say that poverty is increasing, the middle class is shrinking, social mobility is declining, and wealth is becoming more concentrated. We examine data on wealth, income, jobs, mobility, poverty rates, political attitudes, and other factors to see whether the American class system is changing, and if so, how. Among the questions we ask are these: Is the gap between rich and the rest of the population growing? Are opportunities to get ahead better or worse than they were in the past? Are neighborhoods becoming more segregated by social class? Is the balance of political power between classes changing?

The charts in Figure 1.2 preview some of our findings. They tell a story of a remarkable turnaround: Class inequalities, which fell in the 1950s and 1960s, rose steeply after the mid-1970s. This reversal is explicit in the U-shaped curves. Individually, the charts in Figure 1.2 tell us the following about the years since the early 1970s: (a) Wealth is increasingly concentrated in the hands of the richest 1% of households; (b) the income gap between the top 5% and the bottom 40% of families is growing; (c) the proportion of men who work full time, year round, but still have poverty-level incomes has been rising; (d) the country has made no progress against poverty since the early 1970s; but (e) the proportion of families with high incomes (above $100,000) has been growing steadily.

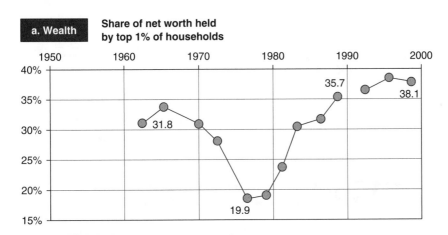

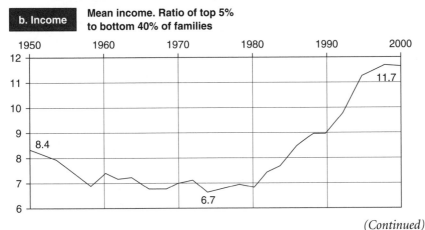

(Continued)

(Continued)

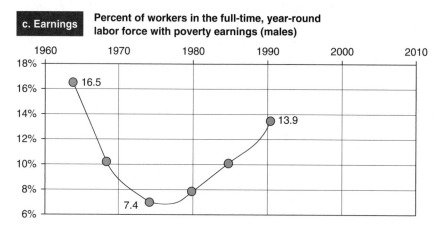

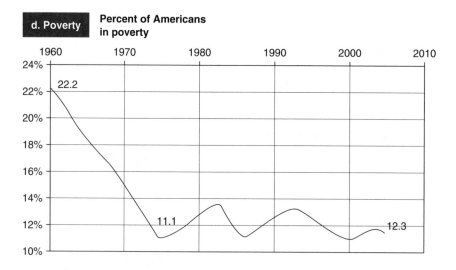

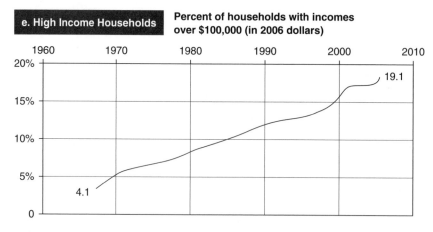

Figure 1.2 From Shared Prosperity to Growing Inequality, 1950–2000

SOURCES: U.S. Census Bureau 1992, 2001a, 2001d, and unpublished tables provided by the U.S. Census Bureau; Wolff 1993, 1998.

We take a second, more careful look at each of these charts in the appropriate chapters. For now, we want to drive home the lesson of what has been called "the great U-turn" (Harrison and Bluestone 1988) and distinguish two periods in recent history. We will call the years after World War II, from 1946 to approximately 1973, the *Age of Shared Prosperity,* and the years since 1973, the *Age of Growing Inequality.*

Conclusion

We close most chapters with a summary of the main points and some general conclusions. That's a little hard to do for this first chapter because it is actually a conceptual summary of the entire book. Much of what we have to say about the American class system and the way we approach the subject are foreshadowed here. Our advice to serious readers is simply to reread this chapter. The effort will be rewarded as you move through the rest of the book.

At the end of each chapter, you will also find a list of the key terms that were used and that are defined in the glossary. The list below is especially long because many basic concepts were introduced in this chapter. We revisit all but a few of them in later chapters.

Key Terms Defined in the Glossary

Age of Growing Inequality	mode of production
Age of Shared Prosperity	occupation
association	power
bourgeoisie	prestige
capital	proletariat
capitalism	social class
capitalist class	social mobility
class consciousness	social status
family	social stratification
Gilbert-Kahl model of the class structure	socialization
household	status
ideology	superstructure
income	underclass
life chances	upper-middle class
lifestyle	wealth
means of production	working class
middle class	working poor

Suggested Readings

Acker, Joan. 2006. *Class Questions: Feminist Answers.* Lanham, MD: Rowman & Littlefield.
   Social class from a feminist perspective.

Bendix, Reinhard and Seymour Martin Lipset, eds. 1966. *Class, Status, and Power: Social Stratification in Comparative Perspective.* 2nd ed. New York: Free Press.

> *A large collection of articles on important aspects of stratification in various countries. Part I covers basic theory, including Marx, Weber, and Davis and Moore's functionalist explanation of stratification.*

Coser, Lewis. 1978. *Masters of Sociological Thought.* 2nd ed. New York: Harcourt Brace Jovanovich.

> *Contains short personal and intellectual biographies of Marx and Weber, which put them into the context of their times.*

Crompton, Rosemary. 1998. *Class and Stratification: An Introduction to Current Debates.* 2nd ed. Cambridge, MA: Polity.

> *Thoughtful guide to recent controversies.*

Grusky, David, ed. 2001. *Social Stratification: Class, Race, and Gender in Sociological Perspective.* 2nd ed. Boulder, CO: Westview.

> *Anthology blending classic and postmodern readings.*

Kingston, Paul W. 2000. *The Classless Society.* Palo Alto, CA: Stanford University Press; and Pakulski, Jan and Malcolm Waters. 1996. *The Death of Class.* Thousand Oaks, CA: Sage.

> *Two provocative books, arguing that the concept of social class no longer corresponds to social reality and should be abandoned by students of social inequality.*

Lenski, Gerhard. 1966. *Power and Privilege: A Theory of Social Stratification.* New York: McGraw-Hill.

> *Ambitious attempt to explain the development of stratification in each of several evolutionary stages, based on technology.*

Marx, Karl. 1979. *The Marx-Engels Reader,* edited by Robert C. Tucker. 2nd ed. New York: Norton.

> *Convenient collection of the writings of Marx and his partner, Friedrich Engels. Particularly relevant are "The German Ideology," Part I; "Wage Labour and Capital"; "Manifesto of the Communist Party"; and Engels's "Socialism: Utopian and Scientific."*

Sernau, Scott and Jonnie Griffin, eds. 2004. *Social Stratification Courses: Syllabi and Instructional Materials,* 5th ed. Washington, DC: American Sociological Association.

> *Varied approaches to the study of stratification. Available from the ASA, 1722 N Street NW, Washington, DC 20036.*

Weber, Max. 1946. *From Max Weber: Essays in Sociology,* edited by H. H. Gerth and C. Wright Mills. New York: Oxford University Press.

> *A selection of Weber's most important sociological writings (except for his book* The Protestant Ethic and the Spirit of Capitalism*). Especially relevant are "Class, Status, Party"; "Bureaucracy"; and "The Protestant Sects and the Spirit of Capitalism."*

CHAPTER 2

Position and Prestige

I am a member of the privileged American class known as the WASPs, the silver spoon people, the people who were handed things from an early age and stepped into a safe, clean, white world.

Thomas Langhorne Phipps

Prejudice against working class people is the last acceptable prejudice. The fact that it's okay to talk about ladies with big hair . . . to talk about trailer trash. Think about these terms applied in any racial or ethnic context.

John DiIulio

I n the small group, in the local community, in the society as a whole, we notice that some people are looked up to, respected, considered people of consequence, and others are thought of as ordinary, unimportant, even lowly. Everywhere we see somebodies and nobodies.

Prestige is a sentiment in the minds of people that is expressed in interpersonal interaction: Deference behavior is expected by one party and granted by another. Obviously, this can occur only when there are values recognized by both parties that define the criteria of superiority (deference at pistol point is not the result of prestige). The parties need not be in perfect accord in their definition of the situation. Deference may be granted gladly or grudgingly. But as long as the subordinate recognizes that the superordinate does have some basis to claim deference and feels constrained by group norms to grant it, a prestige difference exists.

The most direct way to study prestige is to go into a local community where everyone knows everyone else and observe how they treat one another. That observation would be aided by questions about individuals and their standing in the community. As the observer moves from one sphere of local life to another, he or she will soon begin to see a pattern that groups individuals and families into clusters of people who are relatively homogeneous; we call them social strata or prestige classes. Within each, people feel similar and find it easy to communicate and share activities. Between strata, however, there can be some sense of distance and even tension: "They are not our sort of people."

The prestige hierarchy of a local community is based on detailed information about individuals and families, all seen through the lens of the residents' general understanding of the class system. Such exhaustive mutual evaluations are only possible in small communities whose members have extensive knowledge of one another's lives and backgrounds. In other social contexts, people must depend on limited and more superficial clues to assign prestige to others, and the sociologists who study them must invent indirect methods of ranking. Nevertheless, an understanding of such local prestige structures is an excellent beginning for the study of the dynamics of stratification because if we can picture in detail the hierarchy of prestige, we can then look behind it for the factors that created it, both local circumstances and national conditions.

In this chapter, we examine some classic investigations of prestige structures in small towns, a related study conducted more recently in two major cities, and a series of national surveys of the prestige associated with occupations. We particularly ask two questions: How do people create mental maps of the class system from their varied daily experiences of prestige distinctions? and How can the sociologist turn the often vague and contradictory perceptions of respondents into coherent models of the class structure?

W. Lloyd Warner: Prestige Classes in Yankee City

The interest of American social scientists in the prestige aspect of stratification can be traced back to a series of community studies conducted by W. Lloyd Warner and his students and colleagues, beginning in the 1930s. The first of these was in a New England town of about 17,000 that Warner called "Yankee City." Later they conducted studies of even smaller communities in the South and Midwest (Warner and Lunt 1941; Warner et al. 1949b).

Yankee City was once a famous seaport. It had a long history in New England commerce, having been a center of trade, fishing, and, more recently, manufacturing, especially of shoes and silverware. In many ways, its glory was in the past. In recent years, it had become merely a small city not too far from Boston, and many of its young people left for the more exciting life to be found in Boston and New York. Ethnically, the town was relatively homogeneous but not perfectly so. Some families had been there for 300 years. Half its inhabitants had been born in the community, and another quarter came from other parts of New England and the United States. But the remaining quarter was from French Canada, Ireland, Italy, and Eastern Europe.

Warner and his team began their research with the assumption that class distinctions people in Yankee City made among themselves would be determined by economic differences. The initial interviews tended to confirm this view. Their respondents spoke of "the big people with money" and "the little people who are poor." Property owners, bankers, and professionals were high status. Laborers, ditchdiggers, and low-wage workers were low status.

However, after the researchers had been in Yankee City for a while, they began to doubt that social standing could so easily be equated with economic position, for they found that some people were placed higher or lower than their incomes would warrant. They noticed that certain doctors were ranked below others in the social hierarchy, even though they were regarded as better physicians, and that high prestige was associated with certain family names. Such distinctions were often made unconsciously, which made them all the more convincing to the researchers.

Warner discovered a hierarchy of prestige classes in Yankee City, consisting of groups of people who were ranked by others in the community as socially superior or inferior. From his interviews and observations, he concluded that the place of individuals within this system was the result of a combination of economic and social variables that included wealth, income, and occupation but also patterns of interaction, social behavior, and lifestyle. People of the same class tended to spend time together and, as a result, to develop similar attitudes and values. Their children were likely to marry one another. Warner's Yankee City research had, in other words, led him to a conception of social class close to Weber's idea of "status group"—a communal group bound by shared prestige, lifestyle, values, and patterns of association.

Some of the patterns Warner observed were based on kinship. Children were assigned the status of their parents, and certain families had a prestige position that was not entirely explainable by their current wealth or income and seemed to flow from their ancestry.

Warner noted that when a person had an equivalent rank on all the economic and social variables, people in Yankee City had no difficulty determining his or her prestige rank. But when someone had different scores on the several variables, ranking became problematic. This usually meant that the person was mobile and was changing position on one variable at a time. Consequently, time was an important factor in stratification placement. For example, if a man who started as the son of a laborer became successful in business, he would be likely to move to a "better" neighborhood, to join clubs of other business and professional men, and to send his children to college. However, if he himself did not have a college education and polished manners, he would never be fully accepted as a social equal by the businessmen who had Harvard degrees. His son, however, might well gain the full acceptance denied the father.

After several years of study by more than a dozen researchers, during which time 99% of the families in town were classified, Warner declared that there were six groupings distinct enough to be called classes (Warner and Lunt 1941:88):

Upper-Upper Class (1.4%). This group was the old-family elite, based on sufficient wealth to maintain a large house in the best neighborhood, but the wealth had to have been in the family for more than one generation. Generational continuity permitted proper training in basic values and established people as belonging to a lineage.

Lower-Upper Class (1.6%). This group was, on the average, slightly richer than the upper-uppers, but their money was newer, their manners were therefore not quite so polished, and their sense of lineage and security was less pronounced.

Upper-Middle Class (10.2%). Business and professional men and their families who were moderately successful but less affluent than the lower-uppers. Some education and polish were necessary for membership, but lineage was unimportant.

Lower-Middle Class (28.1%). The small businessmen, the schoolteachers, and the foremen in industry. This group tended to have morals that were close to those of Puritan Fundamentalism; they were churchgoers, lodge joiners, and flag wavers.

Upper-Lower Class (32.6%). The solid, respectable laboring people, who kept their houses clean and stayed out of trouble.

Lower-Lower Class (25.2%). The "lulus" or disrespectable and often slovenly people who dug clams and waited for public relief.

Among the notable features of this schema of Yankee City classes are the following: (1) the distinction, at the top, between an old-money elite, the product of New England's long history, and a class of families with more recent fortunes; (2) the distinction between those who work with their hands—members of the bottom two classes, comprising more than half the population—and those who do not, in the higher classes; and (3) the attribution (presumably reflecting what Warner and his associates heard in Yankee City) of moral status to class position—the lower-lowers, are, for example, "disrespectable," while those above them have Puritan morals and are "respectable" or "clean."

Once the general system became clear to him, Warner said, he used clique and association memberships as a shorthand index of prestige position. Thus, among men, there were certain small social clubs that were open only to upper-uppers, the Rotary was primarily upper-middle in membership, the fraternal lodges were lower-middle, and the craft unions were upper-lower. It seems that in cases of doubt, intimate clique interactions were the crucial test: A repeated invitation home to dinner appeared to be, for Warner, the best sign of prestige equality between persons who were not relatives.

Prestige Class as a Concept

Warner maintained that the breaks between all these prestige classes were quite clear-cut, except for that between the lower-middle and the upper-lower. At that level, there was a blurring of distinctions that made placement of borderline families quite difficult. Of course, the placement of mobile families at all levels was difficult.

When Warner said that the distinctions between the classes were clear-cut, he did not mean that people in Yankee City could necessarily give a consistent account of them. Like other Americans, they were uncomfortable with the idea of social inequality. After all, the American creed says that we are born equal. Some Yankee City residents were quite aware of class differences and could describe them. Some denied that classes existed while acting as if they did. Social ranking was often an unconscious process.

But if ranking is unconscious, how can researchers learn about it? The answer, for Warner and his colleagues, was listening to what people said and observing their behavior over an extended period. As this suggests, Warner's version of Yankee City's class system is not a summary of what residents told him about it, nor is it a simple reflection of life in Yankee City. Instead, it is an abstraction from reality, based on systematic questioning, listening, and observation. Warner's analysis is a map of the class system. Like any map, it is a gross simplification of complicated terrain, ignoring irregularities and focusing on what the mapmaker regards as major features.

The individual classes that Warner identified are not mere descriptions of the mental categories used by people in Yankee City and other small towns. They are abstract concepts, designed by the researcher to help organize a vast amount of data on attitudes and behavior. They are not identical with social reality but a useful way of thinking about it. The only sense in which these six Yankee City classes can be called real is to claim that they organize the data better than an alternative set of concepts.

How Many Classes?

Describing the structure of prestige classes in a community is inevitably problematic. The analyst wants to know how many classes there are and where boundaries between them are located but soon discovers that there is little consensus on these matters. One reason, according to a study of a small Southern town by three Warner colleagues, is that the class structure looks different from the perspectives of people at different class levels (Davis, Gardner, and Gardner 1941). Their report, titled *Deep South,* demonstrated this phenomenon with a chart showing how the people at each level perceive the people at other levels. It is reproduced here as Figure 2.1.

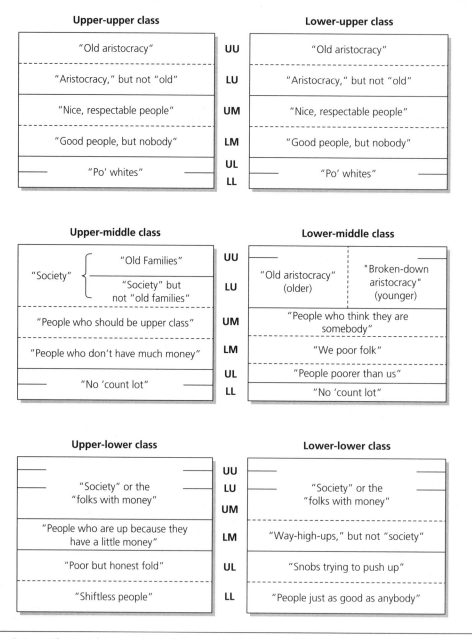

Figure 2.1   The Social Perspectives of the Social Classes

SOURCE: Reprinted from page 65 of *Deep South: A Social-Anthropological Study of Caste and Class,* by Allison Davis, Burleigh B. Gardner, and Mary R. Gardner. Reprinted by permission of University of Chicago Press.

NOTE: UU = upper-upper class; LU = upper-lower class; UM = upper-middle class; LM = lower-middle class; UL = upper-lower class; LL = lower-lower class.

Based on patterns of association and lifestyles, the researchers found six classes among the town's white population, similar to the Yankee City classes. The six boxes in the chart show how the class structure appeared to people in each of these classes. Between the boxes, there are abbreviated labels delineating the classes described by the researchers (UU = upper-upper class, LM = lower-middle class, etc.).

A comparison between the upper-upper class box and the lower-lower class box in the chart reveals large gaps in perceptions. The lower-lowers lump the people in the top three classes into one big class ("Society or folks with money"). Similarly, the upper-uppers collapse the two bottom classes into one ("Po' whites"). People in both classes make more class distinctions at their own level. The labels they use for one another are quite different. For example, the lower-lowers describe the people at the top with phrases suggesting wealth and social pretense, while the upper-uppers describe themselves and nearby classes in terms emphasizing inherited position, social prestige, and respectability.

Some important conclusions about perceptions of the class structure in this Southern town and elsewhere can be drawn from the chart. (Like the chart itself, these points tend to underestimate the differences of perception within classes, while usefully highlighting the differences between classes.)

1. *Number of classes.* People at all class levels perceive class differences, but there is disagreement about the number of classes in the community. No class recognizes a structure of six classes, corresponding to the Warnerian classes. Instead, they see four or five.

2. *Perception and distance.* People make more distinctions among those close to themselves in the hierarchy than among those who are far away. That tendency emerged sharply in the mutual perceptions of the upper-uppers and lower-lowers described in *Deep South*.

3. *Coincidence of cleavages.* Despite class differences in the number of classes perceived at various levels in the hierarchy, the distinctions actually made by people from different classes coincide. For example, the line that the lower-lowers drew between "society" and the "way-high-ups but not society" was the same as the upper-uppers' distinction between "nice respectable people" and "good people but nobody." That is, when the researchers asked people about specific families, they found that the lower-lowers and upper-uppers placed them in the same one of these two groups.

4. *Basis of class distinctions.* People often agree about where individuals or families belong in the class hierarchy, but not about why they are there. In other words, they find different bases for class distinctions. For example, people at the top understand class position in terms of time; they distinguish between "old" families and "new" families. People in the middle make moral evaluations of how things "should be." People in the lower class view the system as a hierarchy of wealth.

From the analyst's viewpoint, perceptions of four or five rather than six classes or discrepancies in the perceived basis of class distinctions are not important as long as the breaks that people do make all fit together in a consistent way. Naturally, people make the finest distinctions regarding those whom they know best and tend to merge others into broader categories. Researchers can take this into consideration and, if they wish, can subdivide a group according to the views of those in and immediately adjacent to it. The task is like that of accommodating for perspective when making a map from aerial photographs. The practical problems in social mapping reduce to two: Do all observers put Albers above Jackson in the hierarchy, and if they distinguish between their ranks at all, do they all divide Jackson's group from Albers's at the same place? These are the questions of ranking consistency and of cutting consistency.

Class Structure of the Metropolis

Several decades after the appearance of the original Yankee City report, two of Warner's former students, Richard Coleman and Lee Rainwater, published the results of their study of prestige classes in two metropolitan areas: Boston and Kansas City. Working in metropolitan areas, Coleman and Rainwater could not duplicate the detailed ethnographic investigation that Warner conducted in Yankee City. Nonetheless, their book *Social Standing in America* (1978) shows the influence of their mentor, to whose memory the book is dedicated.

Social Standing in America was an ambitious undertaking, involving 900 interviews in the two cities. The statistical procedures employed were designed to provide representative samples of adults in Greater Boston and Greater Kansas City. Interviews were standardized and followed a fixed schedule of questions in the style of a social survey, but many questions were open-ended, allowing respondents to describe the class system in their own terms.

The hierarchy of prestige classes that Coleman and Rainwater (1978) stitched together from their analysis of the interviews is rather complex, so we offer a simplified, schematic version in Table 2.1. (Note that the annual incomes were recorded in 1971 dollars. Multiplying these amounts by four will give roughly equivalent values in inflated dollars of recent years.) Inspection of the table shows that the basic structure of the hierarchy is parallel to the one found by Warner in Yankee City. For instance, in both studies, the upper-upper and lower-upper classes correspond to a distinction between established families and "new money," although the distinction might be noticed only by those who are themselves close to the top. In Boston, the upper-class respondents spoke of the former as "the tip-top—as close to an aristocracy as you'll find in America. . . . Yankee families that go way back; the WASPs who were here first . . . the bluebloods with inherited income—they live on stocks and bonds" (p. 150). The same respondents described the lower-uppers as "a mix of highly successful executives, doctors, and lawyers with [very high]

Table 2.1   Coleman and Rainwater's Metropolitan Class Structure

Class	Typical Occupations or Source of Income	Typical Education	Annual Income, 1971	% of miles
I. Upper Americans				
Upper-upper Old rich, aristocratic family name	Inherited wealth	Ivy League college degree; often postgraduate	Over $60,000	
Lower-upper Success elite	Top professionals; senior corporate executives	Good colleges; often postgraduate	Over $60,000	2%
Upper-middle Professional and managerial	Middle professionals and managers	College degree; often postgraduate	$20,000 to $60,000	19%
II. Middle Americans				
Middle class	Lower-level managers; small-business owners; lower-status professionals (pharmacists, teachers); sales and clerical	High school plus some college	$10,000 to $20,000	31
Working class	Higher blue-collar (craftsman, truck drivers); lowest-paid sales and clerical	High school diploma for younger persons	$7,500 to $15,000	35
III. Lower Americans				
Semipoor	Unskilled labor and service	Part high school	$4,500 to $6,000	
The bottom	Often unemployed; Welfare	Primary school	Less than $4,500	13

SOURCE: From *Social Standing in America* by Richard P. Coleman and Lee Rainwater. Reprinted by permission of Basic Books, a member of Perseus Books Group.

NOTE: Percentages adjusted for undersampling of "Lower Americans," acknowledged by authors.

incomes. . . . They have help in the house, fancy cars, frequent and expensive vacations, and at least two houses. . . . They're not considered top society because they don't have the right background—they're newer money, with less tradition in their lifestyle" (p. 151).

In contrast with those at the top, Coleman and Rainwater delineated a bottom class characterized by dependence on irregular, marginal employment or public relief, often shifting from one to the other. As in Yankee City, Boston and Kansas City families in this class were regarded as less than respectable and described in terms suggesting that they were physically and morally "unclean." However, many of Coleman and Rainwater's respondents made a distinction between families on the very bottom and a class of semipoor families who worked more regularly and were slightly more orderly in their lifestyles.

As portrayed by Coleman and Rainwater, then, "upper America" and "lower America" neatly parallel corresponding prestige groupings in Yankee City. The same would appear to be true of "middle America," where Coleman and Rainwater's middle class and working class are equivalent to Warner's lower–middle and upper–lower classes. However, it was in the middle range of the class structure that they had the hardest time organizing the views of Kansas City and Boston respondents into prestige categories. In judging prestige, city respondents at this level gave almost exclusive emphasis to income and standard of living and paid relatively little attention to other stratification variables, such as occupation and association, that had seemed important to Warner.

Coleman and Rainwater reported that their middle-American respondents recognized three levels among themselves, often called "people at the comfortable standard of living," "people just getting along," and "people who aren't lower class but are having a real hard time" (pp. 158–159). But the two sociologists found these categories inadequate and insisted on a more traditional distinction between middle class and working class (each of which, they suggest, can be subdivided along income lines). The distinction the researchers made was essentially between white-collar and blue-collar workers (office workers versus manual workers). They decided to place the lowest-paid white-collar workers in the working-class category. But there are no blue-collar workers, even the best paid, in their middle class. Thus, a highly skilled, well-paid electrician is working class in their schema (see Table 2.1).

How do Coleman and Rainwater justify substituting their own judgment here for their respondents' judgments? They argue that lifestyle and associational differences that emerged in the interviews show that the traditional middle-class/working-class distinction is more fundamental than any income distinction. For instance, among families at the same "comfortable" income level, they noted important differences in consumption patterns. The middle-class families at this level were likely to spend more on living room and dining room furniture and less on the television sets, stereos, and refrigerators and other appliances that were attractive to working-class families with similar incomes. The working-class families owned larger and more expensive automobiles and more trucks, campers, and vans. Moreover, income equality in middle America does not appear to produce social equality: patterns of friendship, organizational membership, and neighborhood location parallel the differences in consumption patterns (pp. 182–183). Some respondents implicitly recognized these differences. A "comfortable" working-class man observed,

> I'm working class because that's my business; I work with my hands. I make good money, so I am higher in the laboring force than many people I know. But birds of a feather flock together. My friends are all hard-working people. . . . We would feel out of place with higher-ups. (P. 184)

The wife of a white-collar man was more explicit:

> I consider myself middle class. My husband works for a construction company in the office. Many of the construction workers make a lot more than he does. But when we have parties at my husband's company, the ones with less education feel out of place and not at ease with the ones with more education. I think of them as working class. (P. 184)

The difficulties Coleman and Rainwater encountered in delineating the prestige hierarchy of middle America raise general, and by now familiar, questions about the methods by which the sociologist defines the structure of prestige classes in a community. They described their approach to prestige measurement as close to Warner's. However, the metropolitan context of their research imposed an important limit on their ability to replicate his methodology. Warner's basic method involved matching the standing of one family against another on the basis of what others said about them and whom they associated with (Warner et al. 1949b). Because the respondents in Boston and Kansas City were unlikely to know or associate with one another, this approach was impractical. Synthesizing individual judgments of the class system was problematic for Coleman and Rainwater because their data consisted of verbal statements about general symbols rather than details about particular others in the community.

Coleman and Rainwater were clear that their version of the prestige hierarchy was not a mirror image of the system as understood in the community. There was, they recognized, no public consensus about the shape of class structure. But they were able to create a composite map of the class structure out of the sometimes inconsistent answers of respondents, each of whom viewed the system from an "inevitably narrow vantage point" (p. 120). Coleman and Rainwater assumed that people were most knowledgeable about the lives of people like themselves; so they gave particular weight to respondents' views concerning the social standing, lifestyles, and associations of those who were near their own level. They listened carefully for repeated references to cleavages in the social hierarchy, which might mark class boundaries.

How does the Coleman-Rainwater metropolitan model differ from the Gilbert-Kahl national model outlined in Chapter 1? (See Table 2.2.) There are some variations in class labels, but the main differences stem from the underlying bases of the models. Our model is based on purely economic considerations, in particular occupation and sources of income. Coleman and Rainwater, like their mentor Warner, have created a prestige model based on public perceptions of the class order, lifestyles, and patterns of association. For this reason, they make the distinction between old money and new money (in effect splitting our capitalist class) and lean toward the traditional blue-collar/white-collar distinction to define the middle and working classes. Despite these differences, the three maps of the class system are broadly similar. This may, in some degree, reflect their common debt to the tradition of sociological thinking about class. But the real key to their similarities is that in Kansas City and Boston, as well as Yankee City, prestige is largely, though not quite wholly, derived from economic position.

Table 2.2   Three Class Models Compared

Gilbert & Kahl (National, 2000)		Coleman & Rainwater (Metropolitan, Early 1970s)		Warner (Small City, 1930s)	
Capitalist	1%	Upper Upper Lower Upper	} 2%	Upper Upper Lower Upper	} 3%
Upper Middle	14	Upper Middle	19	Upper Middle	10
Middle	30	Middle	31	Lower Middle	28
Working	30	Working	35	Upper Lower	33
Working Poor	13	Semipoor	} 13	Lower Lower	25
Underclass	12	The Bottom			
TOTAL	100%		100%		100%

Prestige of Occupation

Warner, Coleman, and Rainwater, and many other investigators, have stressed the importance of occupation for the prestige evaluations Americans make of one another. Especially in metropolitan settings, where people do not have a detailed knowledge of one another's income, family background, lifestyle, associations, and so forth, they are forced to fall back on a few shorthand indicators of personal prestige, such as occupation. They know, of course, that occupation is a fair indicator of two other sources of prestige: income and education. Physicians are typically affluent; not many janitors hold college degrees. People may also associate particular lifestyles and patterns of interaction with occupations or, more generally, with the distinction between blue-collar and white-collar workers. These expectations account for the emphasis they place on occupation in making prestige assessments. For the sociologist engaged in a large-scale research operation, occupation is especially useful: It is more visible than income, and it can be studied with census data as well as social surveys and qualitative field studies. Furthermore, because census data are available for earlier periods, we can use occupation as an indicator in historical research.

There have been numerous studies of occupational prestige, going back more than 50 years, but the best known are the national polls conducted under the auspices of the National Opinion Research Center (NORC) at the University of Chicago since 1947 (Hodge, Treiman, and Rossi 1966; Nakao and Treas 1990; Reiss 1961).

Table 2.3 presents a sampling of occupational prestige scores from a NORC survey. Respondents were asked to rate the "social standing" of each occupation. The scores were created by averaging their responses. Theoretically, the scale runs from 0 to 100, but in practice, scores seldom go above 80 or below 20. The results toward the top and bottom are consistent with what we have seen in the Warner group's community and Coleman and Rainwater metropolitan studies (which, of course, are based on much more than occupation). The highest-ranking occupations are professional and managerial (physician, professor, plant manager, hospital administrator), ordered by the level of expertise or administrative responsibility entailed. Virtually all assume a college education or better.

Table 2.3   NORC Prestige Scores for Selected Occupations

Higher Prestige Jobs

86 Physician
75 Lawyer
74 College professor
74 Computer systems analyst
73 Chemist
72 Dentist
69 Hospital administrator
66 Registered nurse
65 Accountant
64 Public school teacher
62 General manager of manufacturing plant

Medium Prestige Jobs

59 Policeman
57 Construction superintendent
53 Airplane mechanic
51 Electrician
50 Computer operator
48 Manager of a supermarket
46 Secretary
46 Bookkeeper
46 Insurance agent
45 Plumber
43 Bank teller
42 Welder
42 Post office clerk

Lower Prestige Jobs

36 Barber
36 File clerk
35 Assembly-line worker
34 House painter
33 Cashier in supermarket
32 Bus driver
31 Furniture salesperson
30 Carpenter's helper
28 Shoe salesperson
28 Garbage collector
25 Bartender
23 Cleaner, private home
23 Farm worker
22 Janitor
22 Telephone solicitor
21 Filling station attendant

SOURCES: Nakao and Treas 1990, *Prestige Scores for All Occupations*; General Social Survey 2000 files.

The lowest-ranking occupations are unskilled, manual jobs, such as garbage collector, janitor, and filling station attendant. Between these extremes are the less demanding office or sales positions (bookkeeper, secretary, cashier in a supermarket) and the skilled manual jobs (electrician, plumber, welder). But note that there is no clear distinction between white-collar and blue-collar jobs (in Warner's terms, lower-middle and upper-lower). The electrician ranks above the manager of a supermarket, the plumber precedes the post office clerk, and the assembly-line worker outranks the furniture and shoe salespeople. Obviously, when faced with this sort of task, respondents are interested in something more than just where someone works or the color of a shirt collar.

When interviewees in the first NORC survey were asked the main factor they had weighed in making their ratings, the most frequent replies were pay, service to humanity, education, and social prestige, but none of these criteria was volunteered by more than 18% of the sample (NORC 1953:418). Whatever the bases of their judgments, the surveys demonstrate that respondents did have a scale in mind on which they could place occupations with a rough consensus. Although there were significant differences among individuals in their relative ratings of occupations, sociologists were more impressed with the great consistency of the average ratings that were given to occupations by relevant subgroups of the population. The average ratings made by the prosperous and the poor, people in high- and low-prestige occupations, blacks and whites, men and women, residents of the Northeast and the South, and city and country dwellers were almost exactly the same. Even those who proposed different criteria for judging occupations did not differ in the way they ranked occupations. In other words, differences were largely idiosyncratic, reflecting personal views, rather than systemic variations by class, race, gender, or geography.

What systematic differences there were can be summed up in two principles: (1) People tended to raise in rank their own and closely related occupations, and (2) people agreed with each other more concerning occupations that were well known. However, even in the best-known fields, brief occupational titles of the sort employed in the NORC surveys are somewhat ambiguous. Confronted with "lawyer," the respondent does not know if the reference is to a small-town, general-practice lawyer or a partner in a major Wall Street firm. Of course, we can never capture with a survey the richness of detail that Warner reports from a community study because surveys force us to depend on a few simple categories. On the other hand, there is no substitute for the systematic knowledge a survey can provide. It is especially useful for making broad comparisons between different communities or different historical periods, but we must always remember that we are using social symbols about general types of jobs, leaving out a lot of concrete detail.

The consistency of prestige ratings across social subgroups is matched by their stability over time. Researchers noted that the ratings barely changed from one survey to the next, over many decades. The consistency over time and across different social groups suggests that occupational ratings tap a very fundamental dimension of social consciousness.

Occupations and Social Classes

The NORC surveys reveal how the public places occupations on a continuum, but the surveys leave us in the dark about how people might cut that continuum into occupational classes. The problem is obviously similar to the one we raised earlier concerning the grouping of families in Yankee City or Boston into prestige classes. Some evidence on this question was gathered in a national survey of approximately 2,000 adults conducted by the University of Michigan's Survey Research Center (SRC) in 1975 (Jackman 1979). The SRC questionnaire asked respondents to place a series of occupations into one of five class categories: poor, working class, middle class, upper-middle class, and upper class.

The researchers found that people did not have a difficult time associating occupations with social classes (there were few "don't know" responses) and that there was considerable popular agreement about where specific occupations should be placed in the five-class system that was suggested by the interviewer. Majorities of respondents agreed on the class placement most of the occupations presented. But as we can anticipate by now, placing occupations was easier at the top and bottom of the scale than in the middle. There was, for example, strong agreement that high-ranking corporate officers are upper class and janitors and assembly-line workers are working class. However, there was less consensus about where to place a factory foreman or a plumber.

People Like Us

People Like Us: Social Class in America, a recent Public Broadcasting Service documentary (Alvarez and Kolker 1999), probes the ways Americans experience class differences. *People Like Us* is not based on systematic research. But it is provocative and often revealing. It offers glimpses of the raw emotions lingering just beneath the surface when Americans, like those quoted below, talk about class.

Thomas Langhorne Phipps:	I am a member of the privileged American class known as the WASPs, the silver spoon people, the people who were handed things from an early age. . . . We stand better, we walk better, we speak better, we dress better, we eat better, we're smarter, we're more cultured and we treat people better—we're nicer and we're more attractive, and that was built into my sense of who I was growing up. . . . I got a phone call from somebody who decided he wanted to become a WASP . . . [and] would pay me [for] WASP lessons in style. And it was sad because the whole point is . . . *you either are it or you aren't, we believe.* That's the tribal belief, that you either have it or you don't.
Bill Bear, plumber:	I'm standing in line [waiting to pay a bill] and because of the way I'm dressed, I have a tendency to be overlooked, you know? [T]hey really don't want to deal with me. They want to deal with Mr. Suit-and-Tie. . . . I almost started a riot. . . . I had to make it known that I was

next, not Mr. Suit-and-tie. If you want to deal with the son of a bitch, make a date with him later, you know? But *just because I have on working clothes doesn't mean I can't afford to pay my bill, baby.* Well, then she got all embarrassed about that and the manager came out . . . and I just exploded. And uh, there was about four working class guys in there, they all started applauding. Cause they felt the same way. You know? Because society does that automatically.

Barbara Brannen-Newton: I am from the middle class because that's where I was born and that's where I live. Socioeconomically, statistically *we are middle class. But we're black middle class,* and we will always have that word black in front of us until the day I die.

Ginie Polo Sayles: When I was in high school, I went to a country club with a girl-friend. . . . I had never been to a country club and we went swimming in the summer and she said, "Let's go over to the clubhouse and have some fried shrimp and charge it to my daddy." And I'd never had fried shrimp. And I thought, what is that, what will it taste like, what will it look like, will I use the right fork? And what really hit me then was that there were limitations. And that's what I didn't like. *I didn't like the idea of feeling less than, ignorant, eliminated, limited by a class.*

Tammy Crabtree [lives in a trailer that "embarrasses" her teenage son]:   I was on welfare 18 years. And now, I work at Burger King, and I'm trying to make a living, and make a home for the kids. It ain't my fault cause I'm poor. I growed up poor. . . . Even when I'm walkin' to work or something, someone'll hollar, "Ey! Trashy bitch! What're you doing?" I'm just walking to work. All I want is just a life where I can be happy. But right now, I'm not because the way people treats me and the way my kids treat me. . . . *My son, he thinks he's high class and a preppie.* He's the best. He thinks he's better than me, better than his brothers.
(All emphases added.)

People Like Us is often about respect and disrespect, from the blue-blood pride of Thomas Langhorne Phipps to the blue-collar anger of plumber Bill Bear. Barbara Brannen laments that an African American woman can never be fully middle class, however well brought up and educated she may be. Ginie Sayles is intimidated by the country club dining room. Hardworking Tammy Crabtree is made to feel like "trailer trash" by perfect strangers and her own son.

Respect and disrespect is another way of thinking about what we have been calling prestige. While the term *prestige* draws our gaze upward to the world of Thomas Phipps, respect and disrespect broadens our vision to include people in the middle and at the bottom, who may feel injured by class distinctions.

Some of the people who appear in the documentary are trying to claim respect through consumption. Tammy Crabtree's son disapproves of his mother's clothes and is seeking higher status with his own. A man whom the documentary identifies as "a social climber" tells the camera that he would never drive a Ford or even a Volvo; they send the wrong signals. An affluent couple shows us around their elaborately remodeled kitchen, noting the influences of Tuscany, Southern Italy, and "[perhaps] an old French country kitchen or an English Farmhouse."

We tour an upscale kitchenware store with satirist Joe Queenan, who draws our attention to various esoteric cooking implements and a slim bottle mysteriously labeled "Al Sapone di Tartufo Bianco." Those who know what these things are and can afford them, belong here, explains Queenan. But if "you, a working-class person" don't and can't, perhaps you should go to Wal-Mart, where everything is identifiable. The store's customers are defining a lifestyle by surrounding themselves with sophisticated possessions and, at the same time, asserting a superior class position.

A recurring theme of *People Like Us* is that America, in the words of the narrator, is "a nation of tribes." (Theorist Max Weber would call them status communities.) The members of our tribe are the people we live among and feel comfortable with. They share our background. They are members of our social class. The documentary suggests that the boundaries defining tribes are fixed and well defended, but it also provides ample evidence that people move from tribe to tribe.

Americans are famous for their ability to reinvent themselves. By the time we encounter Ginie Sayles, she has married a wealthy man and is wholly at ease in the country club settings that made her feel "ignorant" and "limited" as a teenager. We meet Dana Felty, an ambitious young woman who left her working-class home in rural Kentucky to attend Antioch College and start a career as a journalist in Washington, D.C. Tammy Crabtree's teenage son, growing up in a trailer in rural Ohio, seems to have similar ambitions. The wealthy crowd we see attending a polo game includes people who have made considerable (and probably recent) fortunes in Internet ventures, finance, fashion, and other fields. Perhaps the members of this success elite yearn to join Langhorne Phipps's old-money upper class. Perhaps they couldn't care less. Phipps's people cherish the myth that admission is by birth only. But, as we will see in Chapter 3, the history of the American upper class contradicts this notion.

The tribes of *People Like Us,* like the prestige classes we have been examining in this chapter, are both amorphous and porous. Their indefinite, open character contributes to the ambitions and anxieties chronicled in the film. Americans, the documentary reminds us, are uncomfortable with class distinctions, which seem undemocratic to them. At the same time, they are aware of a prestige hierarchy and may feel pressured to improve their own rank. Unsure of where they stand, they demand respect, cultivate their manners, remake their wardrobes, and remodel their kitchens.

Conclusion: Perception of Rank and Strata

From the studies we have reviewed, three conclusions stand out: (1) In American society, there is a prestige hierarchy of both persons and occupations—this hierarchy or rank order is divided by most citizens into a few categories or classes; (2) there exists considerable ambiguity about just how to define and differentiate them; and (3) there is more agreement about rank order than about the criteria used in making ranking decisions, and more agreement about ranking than about division into classes. To recall our earlier language, there is greater *ranking consistency* than *cutting consistency* among respondents.

Some tentative principles seem to explain the variation we have encountered in perceptions of ranking and grouping:

1. People perceive a rank order.

2. They agree more about the extremes than about the middle of the prestige range.

3. They agree most about the top of the range and make more distinctions about the top than about the bottom. (Perhaps the top is just more conspicuous.)

4. People lump together into large groups those who are furthest from them.

5. People in the middle or at the bottom are more likely to conceive of class differences in financial terms.

6. Those at the top are more conscious of prestige distinctions based on lineage and style of life.

7. Mobility is a source of ambiguity in perception of the prestige order. People find it difficult to "place" mobile individuals. Perception of high rates of mobility leads to the conclusion that class boundaries are amorphous or nonexistent.

These principles connecting social facts with the way people perceive those facts are sufficient to explain why there is no straightforward answer to the question that is asked so often: how many social classes are there in America? The moment we try to answer the question with data that come from the views of ordinary citizens, we are confronted with ambiguities and contradictions. Coleman and Rainwater manage to stitch together a simplified but coherent schema of prestige classes for Boston and Kansas City out of these inconsistent materials. An alternative, taken by the authors of this book and many theorists before them, is to develop a model that does not claim to be based on perceptions of the class order. These alternative models are typically based on economic distinctions rather than prestige rankings. But they must also be fashioned out of inconsistent materials—drawn from history, economic data, and demography. Both approaches involve a combination of art and science. We can expect the analyst to know the facts and make convincing use of them to develop a map of the class system. The map may be more or less effective as a device to interpret social life. But in the final analysis, the class system is not like the solar system, an objective reality we can hope to discover. Our conceptions of it will always be provisional.

Key Terms Defined in the Glossary

association

blue-collar workers

cutting consistency

occupational prestige

ranking consistency

social clique

social strata

socioeconomic status

white-collar workers

Suggested Readings

Aldrich, Nelson W., Jr. 1988. *Old Money: The Mythology of America's Upper Class.* New York: Vintage.

> *Old money and not-so-old money. The ethos of the national upper class explained from within.*

Alvarez, Louis and Andrew Kolker. 1999. *People Like Us: Social Class in America* [Video]. Public Broadcasting Service.

> *How Americans experience social class. A provocative, revealing program, originally broadcast on Public Television.*

Bourdieu, Pierre. 1984. *Distinction: A Social Critique of the Judgement of Taste.* Cambridge, MA: Harvard University.

> *Class cultures and their functions in defining and reproducing class differences.*

Bowser, Benjamin P. 2007. *The Black Middle Class: Social Mobility—and Vulnerability.* Boulder, CO: Lynne Rienner.

> *African American class structure. The history and future of the black middle class.*

Graham, Lawrence Otis. 1999. *Our Kind of People: Inside the Black Upper Class.* New York: HarperPerennial.

> *Perceptive, engaging account by an insider.*

Ossowski, Stanislaw. 1963. *Class Structure in the Social Consciousness.* New York: Free Press.

> *A systematic account of competing basic conceptions of class structure in the history of Western social thought.*

Veblen, Thorstein. 1934. *The Theory of the Leisure Class.* New York: Modern Library. (First published in 1899.)

> *This book introduced the idea that "conspicuous consumption" was the way to buy prestige in competitive America.*

Warner, Lloyd W. et al. 1973. *Yankee City.* New Haven, CT: Yale University Press.

> *Handy abridgment of the entire series of volumes on Yankee City.*

CHAPTER 3

Social Class, Occupation, and Social Change

No business which depends for existence on paying less than living wages to its workers has any right to continue in this country.

Franklin Delano Roosevelt

A post-industrial society, being primarily a technical society, awards place less on the basis of inheritance or property . . . than on education and skills.

Daniel Bell (1976)

I n 1924, sociologists Robert S. Lynd and Helen Merrell Lynd went to Indiana to describe a "typical" American town. For more than a year, the couple studied everyday life in a community of 35,000, Muncie, Indiana, which they called "Middletown." To establish a historical baseline for their work, the Lynds reconstructed life in 1890, when Middletown had only 11,000 people and was going through the first stages of industrialization. They returned in 1935 for a restudy. Thus, we have three points of observation: 1890, 1924, and 1935.

The Lynds' two books, *Middletown* (1929) and *Middletown in Transition* (1937), are classic studies of American life. They are especially valuable to us because they illustrate, in the miniature scale of a small town, how the American class structure was transformed by industrialization. Their narrative links social change, occupation, and social class—our key concerns in this chapter—and provides a historical backdrop against which we can understand the more recent social transformation associated with the emergence of a "postindustrial society."

Middletown: 1890 and 1924

In their first book, the Lynds reported that daily life in Middletown was conditioned by a fundamental social distinction: the division of the population into working class and business class. Generally, members of the first class (70% of the population) supported themselves by working with *things,* while members of the second class (30%) made a living in activities oriented toward *people.* Being born on one side or the other of this cleavage was "the most significant single cultural factor" in the lives of Middletown's inhabitants, influencing matters as varied as who they married, when they got up in the morning, how they dressed, and whether they drove Fords or Buicks and worshiped with the Holy Rollers or the Presbyterians.

One central theme of the first volume was that, from 1890 to 1924, there were basic changes in the work pattern of both the business and working classes—changes, incidentally, that resulted in a wider gap between them. These changes flowed from three causes: larger population, more machinery, and increasing emphasis on money.

Middletown in 1890 was a market town that was just beginning to turn to manufacturing. The work habits and values of its people were extensions of the traditions of their farmer parents—people who had conquered a wilderness. There had been land for all who would work it, and from such plenty, there emerged a society that lacked clear gradations of rank and privilege, a society that stressed individual initiative and progress, family solidarity, simplicity of manners and style of life, and equality among neighbors.

As trading and handicraft manufacturing succeeded farming as the base of livelihood, the old traditions could easily continue. A man earned whatever his own efforts deserved. True, there developed a gradation in income that extended from unskilled through skilled laborers to bosses (who were often former craftsmen) and a few professional people. But income and prestige, which were understood to be direct outcomes of competence at work, tended to rise with age and experience.

The spread of machine production began to change all this. The machine undermined the value of the traditional knowledge and skills of the master craftsman. It demanded speed and endurance. After a few weeks' experience with the machine, a 19-year-old boy could outproduce his 42-year-old father. The old apprentice system was gradually abandoned, and with its passing, the traditional distinction between skilled and unskilled workers was blurred.

There were basic changes among the business class as well. The old businessman was a small merchant or manufacturer, whose capital consisted mostly of his personal savings. He had started as a worker and had become a businessman. His relations with both employees and customers were personal, even intimate. The new businessman operated with bank credit, had too many employees and customers to know them personally, and had ties with other businessmen all over the country. Sometimes he was a branch manager for a national corporation.

Money had become the significant link between people. In the old days, families were more self-sufficient, producing much of their food and clothing at home. When they purchased things, they paid cash. Now they bought most of what they needed, sometimes on credit. The money nexus was becoming increasingly important as more spheres of life became parts of the commercial market. The result was that people began to use money as a sign of accomplishment, a common denominator for prestige.

Within the business class, friendships, social activities, church membership, and political principles were often viewed from the instrumental standpoint of their contribution to one's business success. In turn, social status in these other realms became dependent on financial status. Newcomers to Middletown were judged by material externals—where they lived, what kind of car they drove. "It's perfectly natural," a leading citizen told the Lynds. "You see, they know money, and they don't know you" (1929:80–81).

Middletown Revisited

When the Lynds returned for a restudy in 1935, Middletown had grown larger. Its industrial base was more mechanized, more centralized into larger units, and more subservient to national corporations like General Motors, which owned plants in Middletown. Production was becoming so efficient that an industrial labor force that was only slightly larger could produce vastly more goods. As a result, new employment was flowing into the service sectors of the economy (such as finance, health, education) and the occupational structure was changing accordingly. There were more schoolteachers, more social workers, and more dental hygienists.

At work, the gap between worker and manager had widened. The managers were not so often self-made owner-businessmen but, increasingly, members of "the new middle class" of college-trained professionals and administrators, who worked for a salary.

The factories had become fewer and bigger. They hired more semiskilled machine tenders, fewer highly skilled manual workers, and growing numbers of white-collar specialists, like engineers and chemists. The latter come from outside rather than being promoted from the ranks.

Technological change was transforming Middletown's class structure. It could no longer be adequately described by the simple division into a business class and working class, which had seemed sufficient in 1924. The Lynds described what they saw in 1935 with a schema of six classes, three manual and three nonmanual. Although based on occupational niches, their new conception of the class structure was quite like the one that Warner was describing about the same time for Yankee City, based on prestige, consumption style, and interaction networks.

At the top of the new class structure was an emergent upper class. The Lynds' account of one prominent family of this class, whom they call the "X family," illustrates the process by which new wealth is transformed into style-of-life symbols, interaction networks, personal prestige, and political power. Four brothers founded the family fortune at the end of the nineteenth century by starting a small glass-manufacturing plant on a capital of $7,000. The legends of Middletown contain many tales of the simplicity and humbleness of these business pioneers. When one of the brothers died in 1925, a newspaper recalled that "'he always worked on a level with his employees. He never asked a man to do something he would not do himself'" (1937:75).

These men built an industrial enterprise that became world famous. They grew rich, contributed vast sums to local charities, bought a substantial interest in the local newspaper, and became politically powerful. Believers in high profits and low wages, they led the fight against unionization. The X clan did not run Middletown by itself. There were other powerful business interests, and the people of Middletown still voted. But, concluded the Lynds, "the business class in Middletown runs the city. The nucleus of business-class control is the X family" (1937:77).

Not until the second generation did the X family gain leadership in consumption affairs: Around the adult sons and daughters of the X clan, with their model farms, fine houses, riding clubs, and airplanes, there developed an exclusive younger clan that was self-consciously upper class. These families were set off from lesser business folk by their residence in an exclusive neighborhood and elite patterns of leisure, symbolized by the annual horse show. In contrast, the men of the older generation had always been preoccupied with business and, by and large, disinclined to cultivate an opulent lifestyle.

The second generation of owners is bound to be different: They cannot have the motives of ambitious individuals born in poverty. They are born to the wealth that brings power, the income that brings luxury, and the values of those who have been reared to expect both; these values have been polished by attendance at universities with national prestige, rather than at the local college. They will be bound to seek the company of others like themselves. They will, in other words, create an upper class with links to the national elite.

Industrialization and the Transformation of the National Class Structure

The changes the Lynds chronicled in Middletown reflect the sweeping transformation of the United States into an industrial society. Although the process was underway in the 1840s and 1850s, it was consolidated after the Civil War. Between 1870 and 1929, the national population tripled, but the output of the American economy grew tenfold, largely as the result of the colossal gains in productivity made possible by new industrial technologies. By 1900, the United States was the world's leading industrial nation (L. Hacker, 1970; Ross 1968).

Closely associated with this achievement were other developments of critical significance for the American class system, most of which we have seen played out on a small scale in Middletown. For instance, Americans changed from predominantly rural residents and farmers to city dwellers employed at urban occupations. In 1870, approximately three-quarters of the population lived in rural places (under 2,500 inhabitants), and more than half the labor force was employed in farming. By 1930, most Americans lived in towns and cities, and nearly 80% of the labor force was engaged in nonfarm occupations (Ross 1968:26). During the same period, occupational tasks were becoming increasingly subdivided and specialized. Thus, the steelworker and the metallurgist were among a multitude of successors to the blacksmith.

As the industrial economy expanded, the country's labor requirements grew faster than the native population did. This labor deficit was met by encouraging immigration from abroad, a policy so successful that by 1920, 22% of the industrial labor force was foreign born.

Finally, the period saw a leap in the scale of economic organization. Large corporations emerged as the dominant force in the American economy. Industrial technology requires large enterprises, but it was the drive to control national markets and generate monopoly profits that produced such giants as Standard Oil (whose successors include Exxon), E. I. du Pont de Nemours, and General Electric, through a wave of corporate mergers at the turn of the century (Ross 1968:40–41).

In the next few pages, we will try to show how the class structure was affected by these changes. First, we will look at three broad class groupings (upper class, working class, and middle class) that were decisively transformed by industrialization. Then we will examine how the occupational structure was altered.

The National Upper Class

Our account of the upper class draws on the work of sociologist C. Wright Mills, whose best-known work, *The Power Elite* (1956), is examined in greater detail in Chapter 8. Writing of the history of the upper class in that book, Mills observed that the United States had never had a national aristocracy of the type known in Europe. Without a feudal past or a single national center of wealth, power, and prestige like Paris or London, an enduring "pedigreed" class never developed. But before industrialization, there had been a series of relatively stable and compact regional upper classes across America:

Up the Hudson, there were patroons, proud of their origins, and in Virginia, the planters. In every New England town, there were Puritan shipowners and early industrialists, and in St. Louis, fashionable descendants of French Creoles living off real estate. In Denver, Colorado, there were wealthy gold and silver miners. And in New York City, as Dixon Wecter has put it, there was a class made up of coupon-clippers, sportsmen living off their fathers' accumulation, and a stratum [of families] . . . trying to renounce their commercial origins as quickly as possible. (Mills 1956:48)

The prestige of these regional upper classes rested on old family fortunes, some on the Eastern Seaboard reaching back to colonial times. By gradually assimilating new moneyed families, the established families were able to keep prestige and wealth in close correspondence. But this system was overwhelmed in the decades after the Civil War. The postwar expansion of the industrial economy created unprecedented opportunities for rapid accumulation of new wealth. Before the war, millionaires were rare, but an 1892 survey found more than 4,000 (Mills 1956:101–102). The wealth of the Lynds' "X" family and the fortunes associated with names such as Rockefeller, Carnegie, Morgan, and Vanderbilt originated in this period.

The post–Civil War fortunes destroyed the neat coincidence of family lineage, wealth, and prestige. The established families initially resisted the social pretensions of the new rich, but this was not easy because the size and national scope of the new wealth dwarfed the older family fortunes.

Not by chance, this era produced one of the most famous attempts to define membership in upper-class society, Ward McAllister's "400" list. McAllister, who established himself as an arbiter of New York society on the basis of his close ties to the prestigious Mrs. John Jacob Astor, decided in the 1880s that there were "only about four hundred people in fashionable New York Society. If you go outside that number, you strike people who are either not at ease in a ballroom or else make other people not at ease" (Mills 1956:54).

But McAllister's defensive effort to specify the makeup of upper-class society was obsolete even before his list was published in 1892. A decade earlier, Mrs. Astor had overcome her disdain for Mrs. Vanderbilt's new railroad money and accepted an invitation from her to a fancy-dress ball. Gestures of this sort, Mills argued, have prevented the emergence of an American aristocracy. "Always in America, society based on descent has been either bypassed or bought-out by the new and vulgar rich" (Mills 1956:52).

In the late nineteenth and early twentieth centuries, the family fortunes built after the Civil War were merged with the old society to create a new prestige class, national in scope like the new economy and centered in major cities like New York, Philadelphia, Chicago, and San Francisco. The progress of this merger can be seen in the *Social Register,* an elite directory that first appeared during this era and proved to be the most successful and enduring effort to specify the socially elect. The first *Social Register,* published in New York in 1887, contained 881 families in a careful mix of new and old. Within a few years, registers were issued in other major cities. During the period 1890 to 1920, the rate of admissions to the Social Register was high, reflecting rapid assimilation of the new rich; since then, annual admissions have declined to more modest levels (Baltzell 1958:20).

The Industrial Working Class

The economic transformation that forced a restructuring of the upper class also recast the bottom of the American class structure (T. Brooks 1971; L. Hacker 1970; Link and Cotton 1973). Limited before the Civil War to the mills of New England, factory production spread across the country and created a modern industrial working class. By 1930, a total of 17 million workers were employed (largely in manual occupations) in the sectors most directly affected by industrialization: manufacturing, mining, and utilities and transportation.

Many of the new industrial workers were immigrants—notably Italians, Jews, and Slavs, who came after 1890. These recent immigrants were relegated to the least attractive manual-labor jobs, while native-born Americans, including the offspring of the earlier immigrants, dominated skilled occupations and supervisory positions. Thus, a close association grew up between stratification and ethnicity, with the highest positions being occupied by the members of the most culturally Americanized groups. This system was reinforced by active discrimination against immigrants and their children in schools and politics.

The working conditions and wages offered America's expanding working class were often dismal. In 1897, the average workweek in industry was nearly 60 hours. Long hours and the indifference of employers to safety conditions produced an appalling record of industrial accidents. More than half a million workers were killed, crippled, or seriously injured on the job in 1907 (Link and Cotton 1973:38). The Commission on Industrial Relations, appointed by President Wilson, concluded in 1915 that

> a large part of our industrial population are . . . living in conditions of actual poverty. . . . It is certain that at least one third and possibly one half of the families of wage earners employed in manufacturing and mining earn in the course of the year less than enough to support them in anything like a comfortable and decent condition. (Link and Cotton 1973:38)

The commission discovered that the children of the poor (many of whom were themselves employed in factories and sweatshops) were dying at three times the rate of middle-class children and that 12 to 20% of all children in six major cities were underfed and undernourished.

The early years of the new century saw some improvement in these conditions. State legislatures, over the determined opposition of employers, started to pass legislation controlling safety conditions, providing compensation for injured workers, prohibiting child labor, and regulating the labor of women. As industrial productivity increased, hours of work declined and wages increased. However, the life of the average worker remained unenviable.

The contrast between the conditions of the working class and the opulent lifestyles of the new rich convinced many Americans that class divisions were becoming sharper. This impression was reinforced by the evidence of class conflict between the emerging industrial working class and its capitalist-class employers.

The period from the Civil War to World War I saw the most violent labor confrontations in American history. For example, in 1877, when railroads cut the wages of their already overworked (15 to 18 hours per day) and underpaid workers, a wave of strikes hit the nation's rail system. State militia and federal troops (called out by officials sympathetic to the railroads to "maintain order") provoked bloody confrontations in which scores of people were killed or injured. Millions of dollars of damage was done to company property. In 1892, at Homestead, Pennsylvania, a private army of 300 Pinkerton detectives hired by management fired on striking employees of the Carnegie Steel Corporation. In the ensuing battle, three Pinkertons and seven workers were killed. A few days later, the company prevailed on the governor to send state militia to take over Homestead, which was under the (peaceful) control of a strike committee.

Multiple indictments against strike leaders, charging crimes from murder to "treason against the state of Pennsylvania" broke the strike, and the union itself, by depleting the union treasury with legal expenses. However, not all strikes ended in management victories. When textile workers in Lawrence, Massachusetts, struck in 1912, martial law was immediately declared, and three dozen strikers were arrested and summarily sentenced to prison. After several violent incidents, including an attack by police on a group of women and children from striking families, the strikers gained public sympathy and won concessions from employers.

In the wake of the 1877 rail strikes, E. L. Godkin, the prestigious editor of *The Nation*, wrote, "We have had an uprising not against political oppression or unpopular government but against society itself." These events should disabuse Americans of the illusion that "this was the one country in which there was no proletariat, no dangerous class." Godkin urged reinforcement of the army and militia and reduction of loose talk "about the laborer and his rights" (Litwack 1962:53–56).

If such violent incidents were exceptional, they were nonetheless symptomatic of the era. Workers who sought decent wages and working conditions had to contend with powerful companies, which could generally count on the support of state and federal governments; the courts; and, as Godkin's remarks suggest, the press. Only gradually did the federal government begin to assume a more balanced attitude toward labor conflicts. As late as the 1930s, the legal right of workers to form unions was still in question.

In the late nineteenth and early twentieth centuries, the United States came as close as it ever has to Marx's conception of a capitalist society. Readers of the *Communist Manifesto* would recognize the emergence of a bourgeoisie and a proletariat in an era of rapid industrial development, the creation of a political and ideological superstructure beholden to the bourgeoisie, and evidence of violent class conflict. However, events did not take the course an orthodox Marxist might have anticipated. Despite Godkin's observation about an "uprising against society," American workers proved to be more interested in earning reasonable wages than in creating a socialist society, and they used both their unions and their votes to help achieve their goals. In addition, two tendencies prevented stark polarization between two classes, a development that Marx regarded as a prerequisite to proletarian revolution. One was division within the working class among ethnic or racial groups and among workers at different skill and wage levels. These divisions kept the workers from developing a shared, militant class consciousness. The other was the expansion of the middle class, which offered the chance of advancement to workers or their children. Added to this were long-term gains in the standard of living of most groups that appeared to absorb their attention more than did the differences between rich and poor.

The New Middle Class

For an understanding of the transformation of the middle class in industrializing America, we turn again to the work of C. Wright Mills. In *White Collar* (1951), drawing on the earlier work of Lewis Corey (1935, 1953), Mills distinguished between two groupings he called "old middle class" and "new middle class." The former is composed of small entrepreneurs: farmers, shopkeepers, self-employed professionals, and the like. The hallmark of this class is its independence, based on the entrepreneur's ownership of the property with which he or she works. The new middle class is composed of the salaried white-collar people whose emergence the Lynds chronicled in Middletown—a heterogeneous grouping of office workers, salespeople, and salaried professionals and managers. Early in the nineteenth century, the old middle class, as defined by Mills, comprised as much as 80% of the total population, but as the country's industrial society matured, the working class became the majority and the new middle class increasingly displaced the old. In particular, there were proportionally fewer farmers and many more salaried professionals, salespeople, and office workers.

The changing balance within the middle class, Mills observed, represented the decline of property and the rise of occupation as the principal basis of stratification. In the countryside, many small farmers were forced off the land or became tenants as agricultural prices fell relative to the prices of the urban products that farmers buy. Behind declining prices were the domination of national economic policy by the new industrialists and financiers of the Northeast, intensified competition from foreign agriculture, and, ironically, advances in agricultural technology that enabled fewer farmers to feed an expanding urban population.

In the cities, many businesspeople shared the financial plight of the small farmers. However, in this case the modest relative decline in the proportion of business owners in the labor force masked what was actually happening: Each year, millions of small business enterprises collapsed, only to be replaced by millions of new enterprises, most of them doomed to the same fate. As corporate dominance of the economy expanded, small business was relegated to the highly competitive retail sector of the economy.

If the decline of the old middle class was evident in the fate of the farmer and small-business owner, the emergence of the new middle class was tied to the triumph of the corporation. As the operations of corporations grew in productive efficiency, financial magnitude, and geographic scope, a decreasing proportion of the workforce was required in the actual production of things, but an increasing proportion was needed to manage, design, sell, and keep account of production.

Moreover, the growth of the corporate economy imposed new tasks on government. The expansion of corporate and public bureaucracies meant rapid growth of the new middle class.

Over time, however, as office employment expanded, the social prestige and material advantage conferred by the white collar declined. At the bottom of large bureaucracies, routinized jobs began to approximate assembly-line employment. The wage gap between average white-collar and blue-collar employees shrank. Of course, the better-trained professionals and managers earned salaries well above blue-collar wages. But the line between blue collar and white collar was no longer the critical boundary in the class system, clearly separating working class and middle class.

National Occupational System

The Lynds' chronicle of Middletown and the parallel national accounts by Mills and others suggest that the post–Civil War transformation of the American class system can be viewed as a result of a series of shifts in the distribution of occupations. We know, for instance, that there was a proportional decline in the number of farmers and a proportional increase in the number of factory and office workers. Approaching social change this way has the advantage of being precise—we can count the number of people employed in different occupations at various points in time. It is also directly relevant to the technological and organizational shifts that have reshaped the economy.

The U.S. Census Bureau has been collecting information on the occupations of Americans for over a century, but analyzing this data requires a schema to reduce a shifting constellation of (literally) thousands of occupations to a manageable set of categories. Since the 1940s, the Census Bureau and independent researchers have been using variants of the schema described in Table 3.1, which groups occupations into a set of broad categories. Because this system, invented by a bureau official named Alba Edwards (1943), is widely employed and will be often seen in this book, we take a good look at Edwards's handiwork.

Edwards based his schema on his own estimation of the nature of the work, skill and training, income, and general social standing associated with occupations. He maintained that his categories formed a "socioeconomic" hierarchy of earnings, education, and prestige, ascending together. This is mostly true. But there is some inconsistency, as we often find, in the middle. Edwards seems to have assumed that white-collar jobs consistently outrank blue-collar jobs. In fact, *skilled* blue-collar workers, on average, earn significantly more than clerical workers and rank about the same on surveys of occupational prestige, though they have lower levels of education. Other problems with the Edwards schema come from lumping diverse occupations into broad categories, especially in the upper categories. For example, the professional grouping includes doctors and engineers, who rank high on all three dimensions. However, the most numerous professionals by far are public school teachers, who occupy a more modest position. Another example is the managerial category, which is large enough to encompass the manager of a fast-food restaurant and the CEO of a major corporation.

Table 3.1   Major Occupational Groups, With Examples

Professionals	**Craft, precision production, and repair workers**
Public school teachers	Carpenters
Doctors	Masons
Computer systems analysts	Foremen in construction or manufacturing
Registered nurses	Machinists
Librarians	Telephone repair workers
Engineers	Auto mechanics
Managers and administrators	**Operatives**
Sales managers	Assembly-line workers in manufacturing
Public administrators	Butchers
Bar managers	Gas station attendants
Corporate executives	Truck drivers
Proprietors of small retail establishments	**Service workers**
Technicians (often classified with professionals)	Waiters and waitresses
	Police and firefighters
	Child care workers
Medical laboratory technicians	Maids and janitors
Dental hygienists	Domestic servants
Computer programmers	**Laborers (excludes farm)**
Sales workers	Unskilled construction workers
Retail store clerks	Freight and stock handlers
Insurance agents	Garbage collectors
Manufacturer's sales representatives	**Farm workers**
Clerical workers	Farmers (owner-operator)
Bank tellers	Farm managers
Bookkeepers	Unpaid farm-family laborers
Secretaries	Migrant farm laborers
Data entry operators	

The Edwards schema is less than perfect as a stratification measure. However, it is doubtful that Edwards or his successors at the Census Bureau could have done much better without producing something infinitely more complex. The original purpose of Edwards's schema was to organize occupational data to make broad historical comparisons, and it serves well for that purpose.

The Transformations of the American Occupational Structure

Using Edwards's compilation and more recent census data, we have charted the evolution of the American occupational structure since 1870. The results are assembled in Table 3.2. The occupational upheaval that first transformed Middletown is reflected here: Since 1870, farm occupations have declined from more than half the labor force to less than 3% in recent years. As the percentage of agricultural workers declined, the proportion of operatives, the grouping that includes most factory workers, rose steadily until it became the single largest occupational category in 1950. But after 1950, the proportion of operatives began to decline. This happened despite the steady expansion of manufacturing production. The other two major blue-collar categories, craftsmen and nonfarm laborers, have followed a similar curved trajectory.

Table 3.2   Occupational Structure of the United States: 1870 to 2000

	Percent of the Labor Force							
	1870	*1900*	*1930*	*1950*	*1972*	*1972**	*1990*	*2000*
Occupation:								
Professionals and technicians	3	4	7	9	14	13	17	19
Managers, officials, and proprietors	6	6	7	9	10	9	13	15
Sales workers	{ 4	5	6	7	7	10	12	13
Clerical workers	{	3	9	12	17	16	15	14
Craftsmen and foremen	9	11	13	14	13	13	12	12
Operatives	10	13	16	20	17	16	11	9
Laborers, except farm	9	12	11	7	5	6	4	4
Service workers	6	9	10	10	13	13	13	12
Farmers and farm laborers	53	38	21	12	3	5	3	3
Total	100	100	100	100	100	100	100	100
Number in labor force (millions)	12.9	29.0	48.7	59.0	81.7	82.1	117.9	136.2
Percent of labor force female	15	18	22	28	38	38	45	47

SOURCES: 1870 from Edwards 1943; 1900–1950 from U.S. Census Bureau 1975; 1972 from U.S. Census Bureau, *Statistical Abstract of the United States*: 1973, and U.S. Department of Labor, *Employment and Earnings*, January 1984; 1990 and 2000 from U.S. Department of Labor, *Employment and Earnings*, January 1991 and 2001b.

*The last three columns, based on the 1980 Census classification of occupations, are not directly comparable with the first five, which are consistent with the 1970 Census classification.

While the agricultural categories have continuously contracted and the blue-collar groupings first expanded and then contracted, white-collar employment has grown steadily, as it did in Middletown (see Figure 3.1). By 1980, white-collar workers constituted more than half the labor force. Two classifications within that group have grown especially rapidly in recent decades—professional/technical and clerical. Both having increased their proportion fivefold.

The relatively modest growth of a third white-collar grouping, the managerial occupations, obscures the radical transformation of this classification. The decline of the independent entrepreneur and complementary increase of salaried managers, noted by Mills in the period 1870 to 1940, has not abated. The proportion of managers who were self-employed dropped from more than half in the 1950s to a few percent 50 years later. The United States has truly become a nation of employees.

The service worker category has also changed in composition as it has grown. At the beginning of the century, most of the workers falling under this rubric were domestic servants (or in the more neutral current language of the Census Bureau, "private household workers"). Recently, the largest and most rapidly expanding

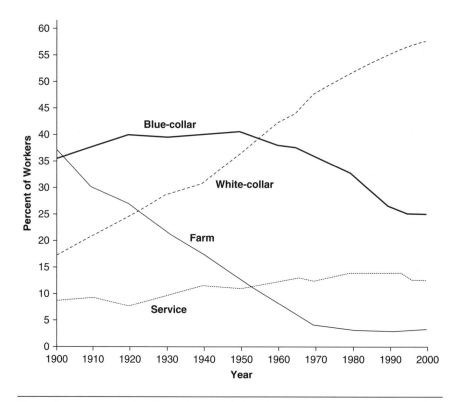

Figure 3.1   Occupational Distribution of the Labor Force

SOURCES: U.S. Census Bureau 1976:387; U.S. Department of Labor, *Employment and Earnings,* January 1990 and 2001b.

service occupations include hospital attendants, waiters and waitresses, janitors, and security guards. While making a hospital bed or washing windows in an office building may be no more satisfying than performing similar tasks in a private home, the change does imply a more democratic set of class relationships. Of course, there are still some privileged American households where domestic chores are done by hired servants. In 2000, there were a little short of 1 million private household workers according to official statistics, but this figure is surely an underestimate, since it misses many household workers who are undocumented immigrants.

From Agricultural to Postindustrial Society

There are several keys to the long-term patterns of occupational change, outlined in Table 3.3 and Figure 3.1. Earlier we noted the growth of large bureaucratic organizations associated with the modern corporation and the expansion of government. This tendency increases the demand for managers; certain professionals, such as accountants; and, above all, clerical workers. The phenomenal expansion of clerical employment is also linked to the increasing participation of women in the labor force.

However, the fundamental cause of changes in the occupational structure since 1870 has been the transformation of the United States from an agricultural to an industrial and finally to a postindustrial society. These three stages are outlined in Figure 3.2. Note that they are defined in terms of the predominant employment sectors of the economy. In 1870, agriculture was still the mainstay of the economy, and most working Americans were farmers. By 1900, the United States had become an industrial society, with most workers employed in manufacturing and related sectors of the economy.[1] After 1970, the United States could be described as a postindustrial society, with most jobs in service-producing rather than goods-producing sectors of the economy. Of course, agricultural production and the output of material goods continued to grow after employment shifted to the service-producing sectors. The United States was producing more food and more industrial goods with fewer workers.

These major economic shifts were conditioned by technological change. The transition to an industrial society was predicated on new technology, which substituted machine power for animal and human muscle. The emergence of a postindustrial society depended on continuing technological advances, which steadily increased productive efficiency and enabled a shrinking proportion of the labor force to meet the material needs of the rest. This dependence on technology, of course, has increased the demand for scientific and technical personnel. Engineering, for instance, is one of the four largest professional occupations.

---

[1]We have abandoned the usual practice of classifying transportation and utilities as service-producing sectors. We include them among the sectors of the industrial stage because they are historically and occupationally closer to the industrial economy (see Figure 3.2).

Periods and Predominant Employment Sectors

Agricultural Society	**Industrial Society**	**Postindustrial Society**
1776–1900	1900–1970	1970–
Agriculture	Manufacturing	Retail and wholesale trade
Fisheries	Mining	Finance and insurance
Forestry	Construction	Real estate
	Transportation	Health care
	Public utilities	Education
	Communications	Government
		Legal services
		Business services
		Social services
		Lodging and restaurant
		Personal services

Figure 3.2   From Agricultural to Postindustrial Society

SOURCE: Based on employment data from U.S. Census Bureau 1975.

As the proportion of the labor force employed in producing goods has declined, employment has expanded in the diverse service-producing sectors, such as retailing, finance, law, hotels, health care, and education. The net result of these changes has been to move employment out of the sectors that require large numbers of blue-collar production workers and into those that depend more on white-collar or service workers. This tendency, as we see later, was reinforced by globalization, which brought growing dependence on imported manufactured goods.

Early assessments of postindustrial society ranged from sunny optimism to dark pessimism. Interpreters on the sunny side anticipated a world of prosperity, opportunity, and social mobility. The postindustrial economy would employ fewer low-skilled manual workers at dirty, back-straining jobs, but require growing armies of professionals and managers. One writer foresaw the "white-collarization of America" (Wattenberg 1974:26). Another expected a "status upheaval" (Bell 1976:134).

Interpreters on the dark side, in contrast, predicted that postindustrial society would provide shrinking opportunities for most Americans and growing social inequality. The sons and daughters of factory workers were more likely to become janitors and food service workers than engineers and computer systems analysts. These writers did not associate "white-collarization" with opportunity but with the declining prestige, shrinking rewards, and increasingly menial character of office work (Braverman 1974; Glenn 1975). Writers on both sides of the debate saw a parallel between the earlier industrial revolution and the contemporary shift toward a postindustrial society. Both anticipated another round of sweeping social transformation in the wake of economic change.

The changes we have examined in the occupational structure (Table 3.2) provide support for both sunny and dark side interpretations of postindustrial society. As the sunny-siders anticipated, the proportion of professionally and technically trained workers has increased rapidly in recent decades and will, according to the Labor Department projections, continue to grow. Detailed statistics show that postindustrial society has required many more electrical engineers, registered nurses, accountants, and computer specialists, but also growing numbers of janitors, hospital attendants, security guards, and food service workers. What is missing from recent trends and current projections are good opportunities for workers with limited skills and education. The kind of factory jobs, especially in heavy industry, that once provided a comfortable and rising standard of living for blue-collar workers and their families are rare in postindustrial society.

Women Workers in Postindustrial Society

A dramatic increase in the proportion of female workers has accompanied the development of postindustrial society. At the beginning of the twentieth century, women constituted only a tiny proportion of the paid labor force. As late as 1950, only 28% of all workers were women, but in recent decades women's labor force participation has accelerated, while the participation of men has begun to decline. By 2000, women constituted almost half the labor force (Table 3.2).

Women workers have long been segregated into a relatively small number of "pink-collar" occupations—fields that are almost entirely female. Secretaries, cashiers, hairdressers, nurses, and elementary school teachers are among the pink-collar workers. Pink-collar jobs typically offer lower pay, less prestige, and slimmer opportunities for advancement than other positions requiring similar levels of education and training. But women workers are increasingly moving out of the pink-collar ghetto.

In the twentieth century, women were drawn into the labor market by the ballooning demand for clerical workers and later for service workers. Beginning in the early 1970s, growing numbers of wives joined the labor force to bolster family incomes that were being eroded by the stagnating wages and the rising unemployment rates of male workers. Increasingly, couples recognized that middle-class and upper-middle class living standards required two incomes.

The pull of economic forces was reinforced by the push of social forces: a declining birth rate, a rising divorce rate, and the increasing proportion of births to single mothers all led women to seek work outside the home. By 1970, the implicit marital compact that had tied men and women to sharply defined family roles in industrial society (husband as permanent provider, wife as permanent housekeeper-caregiver) was collapsing. Attitudes toward working wives and mothers were shifting—encouraged perhaps by the women's movement, but also compelled by the new economic and marital uncertainties.

These changes were particularly felt by the women of the baby boom generation—mothers of today's college students. Compared with their own mothers and grandmothers, the boomers were much closer to men in their educational achievement, late to marry, and more likely to be childless or to delay childbearing. As a result of these factors and others outlined earlier, boomer women were more likely to join the labor force, more career-oriented, more inclined to work full time and continuously, and more like men in their occupations.

For example, in 1979, a woman employed full time, year round, earned, on average, 60 cents for every dollar earned by a similarly employed man. This figure had remained stubbornly unchanged for many years. By 2000, the relative pay rate had risen to 73 cents—and even higher for the baby boomers. In 1970 (the year Hillary Clinton entered Yale Law School), women were 5% of law school graduates; by 2000, they were 48%. During the same period, the proportion of women among medical graduates climbed from 8 to 43%, and the proportion of females among people classified by the Census Bureau as "managers" rose to nearly 50%.

Yet women remain, on average, quite different from men in the marketplace. Even among baby boomers, the majority of women (unlike the majority of men) do not work full time, year round. Their yearly earnings, although advancing, are therefore likely to lag behind men's. As Table 3.3 indicates, women's occupational profile is different from men's. Women are more likely to be employed as clerical or service workers and less likely to hold one of the skilled blue-collar jobs. Although women appear to be well represented in the general category of "professionals," more detailed occupational data indicate that the majority of female professionals work in just three fields: nursing, teaching, and social work. The high-prestige, high-income professions, including law, medicine, engineering, and architecture, continue to be dominated by men.

Table 3.3   Occupational Distribution by Sex, 2002

	Percent of Workers	
	Males	Females
Managers	15.3	14.9
Professionals	13.6	18.9
Technicians	2.8	3.9
Sales	11.4	12.5
Clerical	5.4	22.5
Crafts	18.5	1.9
Operatives	12.9	4.6
Services	10.6	18.1
Laborers	5.9	1.8
Farm	3.8	1.1
Total	100.0	100.0
Number (millions)	72.9	63.5

Thus, the pink-collar phenomenon endures, though it is slowly weakening. By 2000, fewer than half of working women (47%) held pink-collar jobs—if we define a pink-collar occupation as one that is 75% female.

Like the advance in women's relative wage noted earlier, the decline of occupational segregation marks the arrival of the more career-oriented baby boom women. To a lesser extent, it also reflects structural change in the postindustrial economy. Declining employment in sex-segregated occupations drives the index down. The postindustrial decline in blue-collar employment and growth of more integrated white-collar occupations tend to reduce occupational segregation.

Both pink-collar employment and what might be called the Hillary Clinton phenomenon—ambitious, successful women married to ambitious, successful men—are of growing importance for the class system. The reason, as we will see in Chapter 4, is that U.S. households increasingly depend on women's earnings.

Transformation of the Black Occupational Structure

In his classic 1944 report on black America, Gunner Myrdal declared, "The economic situation of the Negroes in America is pathological. Except for a small minority enjoying upper or middle class status, the masses of American Negroes . . . are destitute" (Myrdal 1944, cited in Smith and Welch 1989:519). Even as Myrdal wrote, the situation he described was changing.

From 1940 to 1980, the weekly earnings of the average full-time black worker (in inflation-adjusted "real" dollars) rose 400%. Because white wages were growing at a slower rate, the wages of black men advanced from 43% to 73% of the white male average over the same period and to 80% by 2000.

Table 3.4 reveals the sweeping transformation of the black occupational structure. In 1940, an estimated 80% of black workers were still concentrated in the three lowest occupational categories. In the decades that followed, the black occupational structure shifted decisively toward the higher categories, bringing it closer to the distribution of white workers. The change for black women was even more dramatic than this general picture for both sexes. As late as 1960, one-third of employed black women cleaned white people's houses. Only a small percentage worked at the white-collar jobs that were typical of white women. By the 1980s, about half of black female workers held white-collar positions (Farley 1984:48–49).

Black workers were affected by the same broad processes of socioeconomic change that were altering the world of white workers, but in ways that were peculiar to an oppressed minority. In the 1930s and 1940s, under the pressures of new industrial unions and the needs of economic mobilization for World War II, the discriminatory barriers that had kept blacks out of many industrial jobs began to fall. With the mechanization of Southern agriculture in the 1940s and 1950s, large numbers of black sharecroppers and farm workers were forced off the land, typically into urban slums and low-wage urban employment.

The civil rights movement of the 1960s produced antidiscrimination legislation and affirmative action programs in government and private industry, which opened many new jobs to blacks. At the same time, blacks were closing the

Table 3.4   Occupational Structure of Black Workers, 1940–2002

	Percent of Workers			
	1940	*1960*	*1983*	*2002*
Managers/proprietors	1	2	5	10
Professionals & techs.	3	5	12	15
Sales & clerical	2	8	23	25
Craftsmen & foremen	3	7	9	7
Operatives	10	21	17	13
Laborers	14	14	7	5
Service workers	34	34	25	23
Farm	32	8	3	1
TOTAL	100	100	100	100

educational distance between themselves and whites. As a result, many young blacks were able to qualify for the more attractive white-collar jobs.

Despite the progress since 1940, a considerable occupational gap remains between blacks and whites. Compared with whites, blacks are still underrepresented at the top of the occupational structure and overrepresented at the bottom, most notably among service workers (Table 3.4). There are, to take an extreme case, approximately two white hospital orderlies for every white physician, but twelve black orderlies for every black physician. As this example suggests, the benefits of change have been unevenly distributed among black Americans. In 2002, the proportion of black professionals and managers was more than six times what it had been in 1940, but "service occupations" remained the largest single black occupational category, as it had been 60 years earlier. Furthermore, blacks were much more likely than whites were to have no job at all—especially in periods of economic stagnation.

The net result of these diverse trends has been increasing class differentiation among African Americans. Although large numbers of young, well-educated workers are moving into jobs that were open to few of their parents, the underclass of low-wage or unemployed workers appears to be growing. Income inequality among blacks has been increasing since the mid-1970s.[2] By 1980, sociologist William J. Wilson (1980), an influential scholar of black America, was writing about a "deepening economic schism in the Black community" (p. 142).

Wages in the Age of Growing Inequality

The shift from an industrial to a postindustrial economy roughly coincided with the transition, referred to in Chapter 1, from the Age of Shared Prosperity to the Age of Growing Inequality. We have labeled the quarter century following World

---

[2]For evidence of this trend, see the Gini ratio series for black families available on the Census Bureau Web site. The Gini ratio is commonly used as a measure of income inequality.

War II the *Age of Shared Prosperity* because during these years the annual earnings of American workers at all levels grew at a healthy pace. But, as Figure 3.3 indicates, this pattern of wage growth came to an abrupt halt in the 1970s (Danziger and Gottschalk 1995; Levy and Murnane 1992; U.S. Department of Labor 1994).

We call the period since the mid-1970s the *Age of Growing Inequality.* Three key tendencies characterize job earnings during these years: (1) On average, men's earnings (in inflation-adjusted "real" dollars) have more or less stagnated, in sharp contrast with the growth of preceding decades; (2) women's earnings have risen steadily; and (3) both the distribution of men's earnings and distribution of women's earnings have become more unequal. The first two tendencies, charted in Figure 3.4, have inevitably brought men's and women's wages closer.

The third tendency is substantiated in Table 3.5, which reveals a pattern of growing inequality. Real wages at the bottom and middle of the labor market (10th and 50th percentiles) were hardly different in 2005 from what they had been three decades earlier, but wages at the top (90th percentile) had risen substantially. In short, a widening gulf separates workers in the lower half of the labor market from the high-wage workers at the top. This same tendency exists for both genders, but is much more pronounced among men.

An especially discouraging feature of the new age has been the growing number of men working for what can be considered poverty wages. A 1992 Census Bureau report defined a "low-wage worker" as someone who worked full time, year round, without making enough to maintain a family of four above the federal poverty line ($20,500 at 2005 price levels).[3] Figure 3.5, one of the U-shaped curves we previewed in Chapter 1, shows that the proportion of men in this sad situation dropped in the 1960s only to rise again in the 1980s. Recent hourly wage data indicates that the proportion of men earning poverty wages was 16% in 1979, peaked at 23% in 1989, and then dropped back to 20% in 2005.[4]

At the top of the job market, earnings have soared, especially among the CEOs of major corporations. From 1980 to 2000, the real earnings of CEOs rose by over 600%—a remarkable increase in an era of stagnant wages. The gulf separating the typical CEO from the typical American worker has grown to colossal proportions. The average CEO of a major corporation earned a stupendous 475 times the wage of an average blue-collar worker in 1999, up from roughly 42 times in 1980 (*BusinessWeek*, April, various years).[5] Even among the top executives of the same firm, inequalities of pay have grown. In the 1960s and 1970s, CEOs at major corporations earned 80% more than third-ranking executives at the same company.

---

[3]These low-wage workers would not necessarily be "poor" under the federal poverty standard we will discuss in Chapter 10, since many men in this situation are not supporting families or are supporting families smaller than four, and/or have working wives who also contribute to the family income.

[4]Poverty wage here means an hourly rate at which a full-time, year-long worker could not move a family of four above the poverty line.

[5]The figures are not strictly compatible, but clearly indicative of the widening chasm separating CEOs from typical workers. The 1999 figure referred to "blue-collar workers." The 1980 figure was for "factory workers."

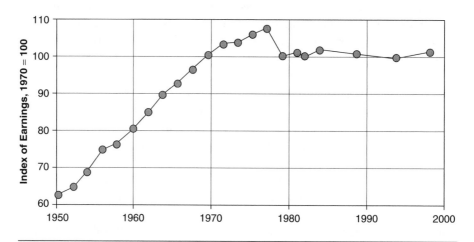

Figure 3.3   Earnings Index for Full-Time Workers

SOURCES: U.S. Census Bureau 1975, *Historical Statistics of the United States,* p. 164 (before 1970), and U.S. Census Bureau, *Statistical Abstract of the United States,* various editions (after 1970).

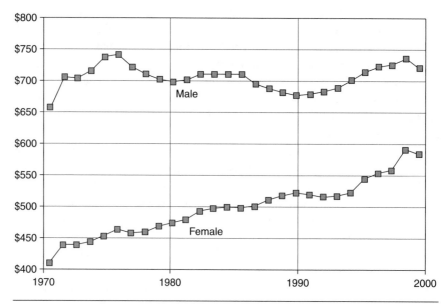

Figure 3.4   Median Weekly Earnings, by Gender

SOURCES: U.S. Census Bureau, *Statistical Abstract of the United States,* various editions.

In recent years, the difference between them has risen to 260% (Porter 2007). The growth of executive pay slowed with declines in the stock market after 2000, but in 2006 chief executives of the biggest 500 corporations earned an average of $15.2 million in total compensation (DeCarlo 2007).

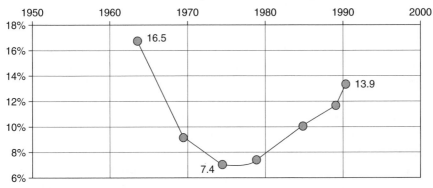

Figure 3.5   Low-Wage Workers

SOURCES: U.S. Census Bureau 1992 and unpublished tables provided by the Census Bureau.

Table 3.5   Wage Inequality, 1973–2005

Percentile/ Percentile Ratio	Real Hourly Wage by Percentile	
	1973	*2005*
Men		
10th	$8.02	$7.79
50th	$15.76	$15.64
90th	$28.83	$36.08
50/10	*1.97*	*2.01*
90/50	*1.83*	*2.31*
90/10	*3.59*	*4.63*
Women		
10th	$5.65	$6.88
50th	$8.83	$11.04
90th	$21.26	$28.86
50/10	*1.56*	*1.60*
90/50	*2.41*	*2.61*
90/10	*3.76*	*4.19*

SOURCE: Calculated from Mishel, Bernstein, and Allegretto 2007:122 and 124.

Growing Inequality of Wages: Why?[6]

The contrast between the Age of Shared Prosperity and the Age of Growing Inequality is epitomized by the fate of a young man, recently graduated from high school, looking for work. A generation ago, he might have found a blue-collar job in manufacturing or construction that would enable him to support a family at a reasonable standard. Today, with the same qualifications, the young graduate would likely find a service or retail position that pays much less in real terms; he would have a tough time supporting a family on his own. Since the late 1970s, the earnings gap between high school– and college-educated workers has widened. So have the gaps between skilled and unskilled and between younger and older workers. Obviously, the tide is running against our high school grad.

In an era of growing inequality, the less educated, the less skilled, and the less experienced have been the biggest relative losers. But the phenomenon of wage polarization does not end with them. At the same time that high school graduates are falling behind college graduates, earnings inequality is increasing among high school grads and among college grads. In fact, such within-group inequality in earnings is rising among the members of virtually any definable group in the labor force: doctors, lawyers, waiters, carpenters, men, women, older workers, younger workers, workers in manufacturing, workers in service industries—even workers in specific firms.

Why is this happening? Four quite plausible explanations have been offered by social scientists:

1. *Economic Restructuring.* As we have seen, in a postindustrial society, employment tends to shift out of manufacturing and into service-producing sectors of the economy (see Figure 3.2). The expanding sectors typically provide ample opportunities for well-educated managers and professionals, but fewer good-paying jobs, like those once available in the goods-producing sectors, for workers with limited education and skills.

2. *Globalization.* In recent years, U.S. participation in international trade has increased. Trade, which generally benefits the U.S. economy, nonetheless creates economic winners and losers in the labor force. Imports of cheap consumer goods produced by low-wage workers in developing countries undercut opportunities for the least skilled American workers. At the same time, increased demand from abroad for high-tech American products from semiconductors to jumbo commercial aircraft creates high-paying jobs for American workers with advanced skills and education. The accompanying expansion of capital markets does the same.

---

[6]This section draws on the following sources: Burtless 1990, 1995; Danziger and Gottschalk 1995; Frank and Cook 1995; Freeman 1996; Gittleman 1994; Harrison and Bluestone 1988; Kodrzycki 1996; Krugman and Lawrence 1994; Levy and Murnane 1992; Mishel, Bernstein, and Schmitt 2001; Sachs and Shatz 1994.

3. *Technological Change.* New technologies, in particular those associated with computers and modern communications, also tend to displace unskilled workers and create increased demand for workers with relevant skills and education. Automation cuts into the blue-collar industrial labor force. The remaining production jobs are likely to require higher skills than they would have in the past.

4. *Weakened Wage-Setting Institutions.* Since the 1970s, "the invisible hand of the market" has gained influence over wages at the expense of institutions like labor unions, "internal labor markets," and minimum wage legislation that previously shielded workers from market forces. Fewer workers are represented by labor unions and covered by collective bargaining agreements, which tend to boost the earnings of unskilled workers and limit wage differentials. Fewer corporations draw on internal labor markets (promoting existing employees rather than hiring outside the firm), a practice that also constrains wage differences. Over time, federal law has allowed inflation to erode the real value of the minimum wage, an important protection for the working poor. (These developments are treated in Chapters 9 and 10.)[7]

There is good evidence for the influence of these factors—although the relative importance of each has been the subject of considerable debate. By now, most thoughtful writers on the subject have concluded that multiple causes are at work. The very pervasiveness of the phenomenon of growing inequality in earnings undercuts simple explanations. For example, the loss of manufacturing jobs has certainly played a role, but it cannot account for the rising within-group inequality among service sector workers or, for that matter, in the manufacturing sector itself.

We conclude this chapter by examining two intriguing interpretations of recent trends: *The Great U-Turn* (1988), by Bennett Harrison and Barry Bluestone, and *The Winner-Take-All Society* (1995), by Robert Frank and Philip Cook. The first looks at developments affecting workers in the bottom half of the labor market, and the second looks at the very top.

Harrison and Bluestone: New Corporate Strategies

American corporations were in serious trouble in the early 1970s, according to Harrison and Bluestone (1988). Labor's share of national income was up. The business share was down. Profits had been falling continuously since the mid-1960s.

---

[7]A fifth category of explanations might be added to this list: labor-market demography. The labor market participation of baby boomers, immigrants (legal and illegal), new college graduates, and women is said to have contributed to wage polarization. Labor supply is inevitably a factor in wage determination, but evidence that change in the supply of workers from these groups has significantly contributed to change in the distribution of earnings is not strong.

Corporate executives were contending with a demanding workforce, abrupt increases in the price of energy, rising taxation, growing government regulation, and accelerating inflation. But the central problem facing American business was rising competition from abroad. Producers in Europe, Japan, and newly industrializing countries such as South Korea and Taiwan were claiming large shares of the U.S. market for products from shoes and textiles to steel and automobiles.

According to Harrison and Bluestone, American capitalists responded to this challenge on three fronts: They adopted new strategies to reduce labor costs, they pursued short-term profit through speculative financial dealings, and they sought favorable government policy in areas from taxes to workplace safety rules. The authors suggest that all three have contributed to increasing inequality, but they particularly stress corporate labor strategies.

Corporations shrank their labor costs, say Harrison and Bluestone, by becoming lean and mean—that is, by reducing the number of people they employ and cutting the compensation of those who remain. Massive layoffs by major corporations, so-called "downsizing," were commonplace in the 1980s and remained so in the 1990s. Industrial corporations took advantage of cheap labor abroad by opening plants in developing countries. Another strategy was "outsourcing," buying components or finished products from low-wage companies at home and abroad. To reduce the cost of labor, companies imposed wage freezes (which depend on inflation to cut real wages over time), reduced benefits, and adopted "two-tier" wage systems providing lower pay scales for new hires. After downsizing their permanent workforces, corporations increased their use of part-time, temporary, and "leased" employees, who typically worked at lower wages, often without benefits. Corporations also found ways to reduce, if not escape, the influence of labor unions. Often, unionized plants were closed and replaced by nonunion plants in regions of the United States or foreign countries unfriendly to union activity. The very threat of such action could be used to intimidate union members and their organizations.

Harrison and Bluestone's account incorporates most of the explanatory factors we listed earlier. In an analysis relevant to the first factor, they measured the effect of replacing manufacturing jobs with service sector jobs. They found that about one-fifth of the increase in wage inequality can be attributed to these job shifts and four-fifths to the changing wage distribution within sectors of the economy. In other words, the postindustrial shift toward services is important, but not as important as the growing wage inequality affecting all sectors. Their description of corporate strategies to hold down wage costs helps explain these broader developments. On the other hand, the authors have little or nothing to say about the role of technology in reducing opportunities for less-educated workers.

Unlike many writers on such topics, Harrison and Bluestone do not simply attribute change to impersonal economic forces. They see corporate leaders making conscious decisions that have adverse consequences for their workers. They argue that there were alternatives: Instead of laying off workers, squeezing wages, and focusing on short-term profit, corporations could have saved good jobs by investing in new technology, modernizing their plants, and retraining their employees.

Frank and Cook: Winner Take All

What do Placido Domingo, Alex "A-Rod" Rodriguez, Ray Irani, Tom Hanks, and Danielle Steel, have in common? They all compete successfully in what Frank and Cook (1995) call "winner-take-all markets"—fields in which rewards are heavily concentrated in the hands of a few top performers. The key to their extraordinary success is relative rather than absolute ability: They are a little better than the competition. For example, probably hundreds of tenors in the world are almost as talented as Placido Domingo, but they will never play major opera houses, get lucrative recording contracts, or appear on television. Like the hundreds of actors who are almost as talented as Tom Hanks, these tenors will be lucky if they can even support themselves as performers.

Frank and Cook assert that the winner-take-all phenomenon, well established in the entertainment industry and in professional sports, is rapidly spreading to other fields, such as business, law, journalism, medicine, and academia. They point to executives like Ray Irani, CEO of Occidental Petroleum, whose total compensation in 2006 came to $322 million. That year, the top 10 CEOs all earned in excess of $95 million (DeCarlo 2007). These are obviously extreme cases, but as we have seen, corporate CEOs now routinely earn millions each year. A yawning chasm separates their compensation from the rewards available to their own employees—even other ranking executives of the same firms. For reasons we do not have space to describe here, the authors are convinced that winner-take-all competition has negative consequences for individuals and the economy as a whole.

The key question for Frank and Cook is, why are winner-take-all markets proliferating? Some writers point to the cozy relationships between executives like Irani and the corporate boards that set their compensation. Others emphasize the emergence in the 1980s of "a culture of greed," tolerant of excessive rewards for those at the top. Frank and Cook stress technological and economic factors. They note that mass production, large-scale organization, and vast markets favor large relative rewards at the top. For example, few authors can claim the millions Danielle Steel gets for one of her (less than profound) novels. But a publisher who commits large sums to producing and promoting mass-market fiction might be ill-advised to sign a writer who is *almost* as popular as Steel. The largest corporations operate on a scale that even 20 years ago would have seemed extraordinary.

With so much money riding on every decision, boards of directors are not inclined to hire someone *almost* as able as Irani for a lot less. Improvements in communications and transportation, along with growing international trade, expand the arena in which winner-take-all markets can flourish. Yet, they are by no means universal. Frank and Cook cite evidence that the large relative rewards flowing to CEOs in the United States are exceptional. The pay gap between European and Japanese chief executives and their workers is modest by American standards. The authors suggest that what makes the United States different is the open competition for top slots in American corporations. In Europe and Japan—as in the United States until recently—executives typically spend their entire careers with a single firm and are promoted from within. Their corporate employers are not compelled to bid against one another for CEOs.

Much of Frank and Cook's argument rests on the idea that barriers to market competition have been falling. Corporations, sports teams, TV networks, universities, and other well-financed organizations are much more willing to raid one another's talent than they have been in the past. Federal deregulation of commercial aviation, trucking, banking, communications, the securities industry, and other sectors increases competition between firms and raises the bidding for top managers and professionals. Falling barriers to international trade and investment probably have the same effect.

Frank and Cook illuminate some of the market forces that are polarizing earnings, but their central concept of winner-take-all remains problematic. The notion works well enough for opera singers and CEOs, but how much does it tell us about the more general phenomenon of rising inequality? The authors attempt to extend the idea to people they call "minor-league superstars"—successful doctors, dentists, lawyers, stockbrokers, accountants, and others who earn hundreds of thousands of dollars a year. They show that inequality of earnings among people in these occupations has been growing. But winner-take-all assumes that a few top players suck up a large share of the available rewards, leaving little for the less talented. It is difficult to imagine that this is the case among people in large, varied professions like law and accounting. On the other hand, the singers, CEOs, and doctors do have this much in common: growing competition has raised the stakes and widened the compensation in their fields.

Another problem is this: Frank and Cook focus entirely on occupational earnings, as we have generally done in this chapter. But, as we will learn in the next chapter where we broaden our perspective on inequality, those with the highest incomes typically depend more on investments than on jobs or professions.

Conclusion

This chapter emphasized the evolution of the occupational structure and its implications for the class system. We have seen the country transformed from an agricultural society; to an industrial society, built around the urban, goods-producing sectors of the economy; and finally to a postindustrial society, dominated by service-producing sectors from fast food to health care. (These shifts are summarized in Figure 3.2.) In the transition to an industrial society, the farm population dropped precipitously, while in the cities, a proletariat of industrial workers grew and a stratum of white-collar office workers appeared. During these years, a national capitalist class emerged.

As industrial society gave way to postindustrial society, farm employment was reduced to numerical insignificance, the industrial proletariat contracted, the white-collar sector swelled and diversified, and a new stratum of service-oriented menial workers emerged. Both industrial and postindustrial society saw the displacement of independent entrepreneurs by salaried employees.

Postindustrial society drew what we (half facetiously) described as sunny- and dark-side interpretations. The sunny interpretations highlight opportunities for well-trained professionals, technicians, and managers in the postindustrial economy.

The dark interpretations emphasize the loss of good-paying blue-collar jobs associated with the decline of manufacturing. For the sunny-siders, the emerging service-producing sectors mean new jobs for hospital administrators, medical technicians, accountants, financial analysts, hotel managers, and computer specialists. For the dark-siders, the service-producing sectors mean employment in lowly service occupations for fast-food workers, janitors, and hospital orderlies. Occupational trend data gives some support to both positions, suggesting an occupational polarization.

Recent trends in occupational structure and earnings distribution support the idea we introduced in Chapter 1 of a shift from an Age of Shared Prosperity (1946–1973) to the current Age of Growing Inequality (1973–). Although this transition roughly coincides with the transition from industrial to postindustrial society, we do not mean to identify one with the other. The developments we associate with postindustrial society have probably contributed to rising inequality, but they may be less important than other sources of inequality. On the other hand, we cannot say that inequality will inevitably continue to grow under postindustrial circumstances. Similarly, the years of industrial society (1900–1970) were not always years of shared prosperity.

The distribution of job earnings gives us a fairly precise way to measure the growing inequalities associated with economic change. The data reveal increased wage polarization, especially among men. At the top of the distribution, earnings have climbed rapidly—in the case of corporate CEOs, spectacularly. At the bottom, earnings have stagnated or even fallen.

The last sections of this chapter explored the possible reasons for the rising wage inequality, including economic restructuring; expanding trade; changing technology; and weakened wage-setting institutions, such as union and minimum wage protections. From Harrison and Bluestone *(The Great U-Turn),* we learned how business strategies designed to cut a corporation's total wage bill have contributed to the declining fortunes of blue-collar workers. Frank and Cook *(The Winner-Take-All Society)* showed how increased market competition has reinforced the concentration of earnings at the very top.

"Earnings" refers to the part of income from jobs, professional practices, and other self-employment. In the next chapter, we look at all sources of income, and instead of focusing on individual workers, we consider entire households. From this broadened perspective, we reexamine the question of growing inequality.

Key Terms Defined in the Glossary

Age of Growing Inequality
Age of Shared Prosperity
agricultural society
blue-collar workers
chief executive officer
downsizing
earnings
industrial society

minimum wage
new middle class/old middle class
occupation
occupational structure
operatives
outsourcing
pink-collar occupations
postindustrial society

(Continued)

service sectors
service workers
two-tier wage systems

wage-setting institutions
white-collar workers
winner-take-all markets

Suggested Readings

Bell, Daniel. 1976. *The Coming of Post-Industrial Society.* New York: Basic Books.
    Broad and optimistic interpretation of postindustrial society, emphasizing changes in technology, economic organization, and occupational structure.

Birmingham, Stephen. 1987. *American's Secret Aristocracy.* Boston: Little Brown.
    A charming, anecdotal account of the world of America's oldest elite families.

Braverman, Harry. 1974. *Labor and Monopoly Capital.* New York: Monthly Review Press.
    Key work on the division of labor and the transformation of work in capitalist industrial societies. Challenge to Bell.

De Luca, Rita Caccamo. 2001. *Back to Middletown: Three Generations of Sociological Reflections.* Palo Alto, CA: Stanford University Press.
    Middletown revisited, this time by an Italian social scientist.

Ehrenreich, Barbara. 2001. *Nickel and Dimed: On (Not) Getting by in America.* New York: Henry Holt.
    Sociologist Ehrenreich took a series of low-wage jobs and tried to live on what she earned. An engaging, revealing account of her experience.

Faux, Jeff. 2006. *The Global Class War: How America's Bipartisan Elite Lost Our Future and What It Will Take to Win It Back.* New York: John Wiley.
    Globalization and the shifting balance of power between labor and capital.

Frieden, Jeffry A. 2006. *Global Capitalism: Its Fall and Rise in the Twentieth Century.* New York: Norton.
    The second half of this well-informed, gracefully written book covers the global trends that shaped the Age of Shared Prosperity and Age Growing Inequality in the United States.

Harrison, Bennett and Barry Bluestone. 1988. *The Great U-Turn: The Corporate Restructuring and the Polarizing of America.* New York: Basic Books.
    The transformation of the American economy in the 1980s and its distributive impact.

Lassiter, Luke Eric et al. 2004. *The Other Side of Middletown: Exploring Muncie's African American Community.* Walnut Creek, CA: Altamira Press.
    Fills a gap in the Lynds' account of Middletown.

McCall, Leslie. 2001. *Complex Inequality: Gender, Class, and Race in the New Economy.* New York: Routledge.
    How gender, class, and race shape the effects of economic restructuring on people's lives.

Mishel, Lawrence, Jared Bernstein, and Sylvia Allegretto. 2007. *The State of Working America, 2006/2007.* Ithaca, NY: Cornell University Press.
    A wealth of information on trends in jobs, wages, and related topics. Updated editions published regularly.

Terkel, Studs. 1974. *Working: People Talk About What They Do All Day and How They Feel About It.* New York: Avon.

   A provocative series of interviews with people, from ranking executives to unskilled laborers.

Wilson, William J. 1980. *The Declining Significance of Race.* 2nd ed. Chicago: University of Chicago Press.

   Argues that changes in occupational structure may be more important for blacks than are traditional forms of prejudice.

CHAPTER 4

Wealth and Income

Whoever said money can't buy happiness simply didn't know where to shop.

Bo Derek

P rocrustes, a giant of Greek mythology, had the bizarre habit of altering the stature of his house guests to fit the length of the available bed by either stretching them or chopping inches off their legs. In the next few pages, we apply Procrustes' approach to the study of income. We have put together an imaginary parade in which the heights of the marchers are made proportional to their incomes (Pen 1971:48–59). The parade is a convenient way of gaining an overview of the distribution of income, our first concern in this chapter. Later we consider the distribution of wealth, and the trend toward greater inequality in the distribution of wealth and income.[1]

Income can be defined as monetary gain over a specified period of time—for example, $50,000 per year. (It should be distinguished from wealth, which is recorded at a *point in time*.) Job earnings, which we examined in the last chapter, are one source of income, but as we will see, there are several other important sources.

The Income Parade

The procession is organized as follows: All the 114 million households counted by the Census Bureau will be represented in the parade. In good Procrustean fashion, the marchers will be stretched or trimmed in proportion to their household's total income. Those representing households with the average annual income ($63,300) will be of average height. By this standard, a marcher from a $100,000 household would be almost 10 feet tall. A marcher from a $30,000 household would be about 3 feet tall.

Because we want a quick impression of the distribution of income, we make the entire procession pass by our reviewing stand at a uniform pace, in exactly 1 hour. This will be rough on the marchers, but it has a particular advantage for us. At any moment, we are able to tell how much of the parade has gone by and how much is to come just by looking at our watches. Let's begin the parade with the shortest marchers, the income pygmies, and work up to the giants. (For an overview of the parade, see Figure 4.1.)

The procession opens on an odd note. In the first few seconds, we see nothing except a few wisps of hair moving across the horizon. It seems that the first people are marching in a deep ditch. They do not appear above ground because they are business people who have suffered income losses during the past year and have had to borrow from the bank or use up their own capital to cover them. Given the high failure rate of small businesses, we should not be surprised at this sight, however peculiar.

Five Minutes: Poor. Next, we see the Tiny People, the size of a match or a cigarette. Five minutes into the parade, the marchers are still only 1 foot tall because they

---

[1]To create the income parade, I have depended on statistics from the Census Bureau's Current Population Survey for 2005, available on the Bureau's Web site, along with Internal Revenue Service (IRS) 2007, Rose 2007, and U.S. Department of Labor 2001a.

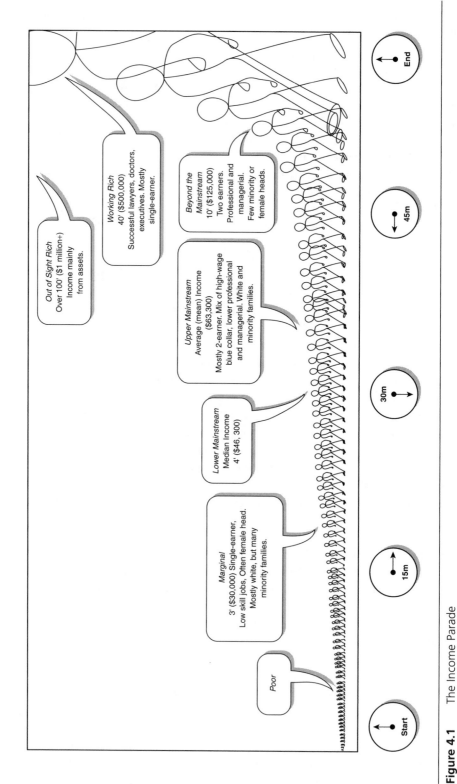

Figure 4.1 The Income Parade

70

receive about $10,000 a year. All who have gone by so far (and many who are to come, depending on the size of their families) are "poor" by the federal government's official poverty standard. There is a notable overrepresentation of women among these dwarfs. Many are female heads of families who are single, divorced, or separated. Even more are elderly women living alone. Close to half of the marchers at this early point in the parade receive income from government transfer programs, such as public assistance, social security, disability payments, or veterans benefits. Some of the marchers are malnourished because their food needs were not reduced when we shrank them to their present size. They might make good use of government food stamps, but many of those who qualify do not receive stamps, either because they are too proud to accept them or because they are ignorant of their rights.

Blacks and Hispanics show up in disproportionate numbers in the first part of the parade. Nevertheless, the majority of the dwarfs who lead the procession are white and non-Hispanic. Actually, the single most encompassing characteristic of the dwarf marchers is that they are not employed because they are old or disabled, temporarily out of work, studying, home with children, or not especially interested in working. Yet, a substantial minority does work, and among them are many people who work full time, all year long, without exceeding dwarf height; they simply are not paid much for their labor.

Twenty Minutes: On the Margin of the Mainstream. As the procession moves on, the marchers get taller, but only very gradually, despite the breakneck pace we have imposed. After 20 minutes, one-third of the parade has passed—we are still seeing 3-foot Little People who live on $30,000 a year. As their not-quite-normal height suggests, the marchers are on the lower margin of the broad mainstream—above the official poverty line ($20,000 for a family of four) but well below the average household income.

Many of the marchers at this point in the parade represent female-headed households. Some are women with children at home; others are women living by themselves. There are many retired people. But the typical household has one wage earner, who works at a low-skilled blue-collar, clerical, or service job. They are assembly-line workers, drivers, sales clerks, janitors, secretaries, and waitresses. (The jobs we refer to in this section are the occupations of the principal income earner in the household.)

What sort of lifestyle do these Little People buy with their money? The answer depends on factors including household size and stage of life. Families of three or four lead austere, sometimes precarious lives. Smaller households can enjoy greater comfort and security. Retirees may benefit from owning homes, free of both rent and mortgage obligations. But typical families at this point in the parade cannot afford to buy a home. If they do own, they are likely to be burdened with high-payment, high-interest mortgages that strain their budgets and contribute to financial instability. They own cars, most often older models, purchased used. Transportation (including recent high gasoline prices), food, housing, utilities, and taxes consume most of the household budget, so there is little money available for other necessities such as clothing. Because these households have little or no savings, even a few weeks of unemployment or an unexpected bill can threaten their standard of living.

Thirty to Forty Minutes: In the Mainstream. At exactly half past the hour, we catch sight of the Midgets, slightly over 4 feet tall, who receive the median income of $46,300. (The median, by definition, is the amount dividing the bottom 50% from the top 50% in any distribution.) A few minutes later, we notice marchers whom we can look in the eyes, assuming, of course, that we ourselves receive the mathematical average (mean) income of $63,300 and are therefore of average height. A careful count reveals that minorities are only slightly underrepresented in this part of the parade, but we do not often see female-headed households.

Many of these average-sized marchers have lower-level managerial or professional positions. Others are technicians or skilled blue-collar workers. But most of the households represented here depend on the earnings of two or more workers. This is especially true of the black and Hispanic families marching in this part of the parade.

These average marchers live substantially better than the Little People we saw 15 minutes ago. They are more likely to own homes and drive late-model cars. Beyond the basics of food, housing, utilities, and transportation, there is room in the budget for other necessities and some luxuries, such as family vacations. Nonetheless, they often feel financially pressed and are not much more likely than the Little People to have money left over at the end of the year.

Fifty to Fifty-Five Minutes: Beyond the Mainstream. Nearing the end of the parade now, we see marchers who would fascinate an NBA scout: They are lanky 10- to 12-footers. The "Lankies" are beyond the mainstream but not quite rich. Their $100,000 to $125,000 incomes allow them to live more gracefully and comfortably than the smaller people who went before. There is money for fashionable clothing, new cars, quality furniture, and perhaps some domestic help at home.

The Lankies typically hold professional and managerial jobs. But few households attain Lanky status with one good job. They are even more likely than the average-sized people to depend on the earnings of two working spouses. Female-headed families are rarely found here. Minority marchers have not vanished from the parade, but their ranks have thinned out in the last 10 minutes.

The Final Minutes: The Rich. The procession has less than 2 minutes to run. Yet, some of the most extraordinary moments lie ahead. If we look down the line at the people who have yet to pass, it appears as if a steep mountain peak is advancing on our reviewing stand. In the final seconds of the parade, we will see, in quick succession, 75-foot Super Giants; 200-foot Leviathans; and, finally, the Big Toes, who flit by in the last fraction of a second. Who are the Big Toes? People like Apple CEO Steve Jobs, Texas investor T. Boone Pickens, and hedge fund manager Steven A. Cohen, all with annual incomes in the hundreds of millions or even billions of dollars. Their big toes, proportional to their towering incomes, are the size of office buildings.

At the end of the parade, the character of the marchers is changing rapidly. The Lankies we saw a few minutes ago depend on two incomes. But Giant households and those that follow do not typically include a working wife. The Giants are likely to be members of the group we call the working rich because their incomes, now in the hundreds of thousands, would drop sharply if they stopped working. They are generally highly successful professionals (most often lawyers and doctors), certain sales people (for example, stock brokers), mid-ranking corporate executives, or the owners of prosperous small enterprises.

Another shift comes with the arrival of the Leviathans and the Big Toes, all with incomes over $1 million. About 250,000 households have incomes in this category. Many of these people hold important jobs; among them, for example, are the top officers of large corporations. However, the greater part of income at this point in the parade does not come from jobs. These marchers own substantial business enterprises, commercial real estate, and valuable portfolios of stocks and bonds. Such income-producing assets, rather than salaries, account for their colossal incomes and overpowering stature. Unlike their poorer cousins, the Giant working rich, their incomes do not depend on reporting to work every morning.

How do these lofty marchers spend their money? A typical urban-based family with a $1 million income owns two homes—a $2 million to $4 million apartment in the city and a substantial weekend house in the country. In addition to housing expenses, the family's annual budget includes $100,000 for domestics (including a nanny, if needed); $40,000 for private schools; and $100,000 for daily expenses, including food. This budget might sound modest to the wealthy family (net worth: $50 million) described in a recent book on the rich (Frank 2007:149). Among the family's expenses were the following:

Mortgages on two homes: $400,000

Domestics and personal assistants: $500,000

Gardening and pool maintenance: $140,000

Cars: $300,000

Air charters: $350,000

Club memberships: $225,000

Charities: $500,000

Political contributions: $61,000

Lessons From the Parade

This chapter elaborates on some of the themes introduced by the procession.

But before going on, we should pause to review what we have just seen and list the general lessons that can be drawn from the parade.

1. *Many Little People, Few Giants.* Our most general impression, confirmed by Figure 4.1, is an extremely gradual increase in income levels until a break point late

in the procession. At half past the hour, we were still looking at people who are 4 feet tall. The slow climb continued until the final minutes of the parade, when heights rose precipitously as the small numbers of Americans with very high incomes, and finally colossal incomes, strode by. According to the IRS (2007), only 2% of tax returns show incomes over $200,000. Only a tiny fraction of a percent exceed $1 million.

2. *Living Standards.* The parade tells us something about the relative welfare of different segments of the population. It took about 20 minutes before we caught sight of the $30,000 Little People and 50 minutes before the $100,000 Lankies appeared. The parade was in its last seconds when we saw the $500,000 Giants. We saw that many families at the $30,000 level could not afford to own a home, most families at the $100,000 level were quite comfortably housed, and families earning $750,000 were likely to own two luxury residences.

3. *Job(s).* The number of workers in each household was one of the main determinants of its place in line. At the beginning of the parade, we noted that many households had no job income. The Little People households that followed typically had one wage earner. Mainstream households usually depended on two workers. At the end of the parade, affluent households with just one income earner were again typical.

4. *Sources of Income.* Jobs are the main source of income for most households. However, during the parade, we noted shifts in the relative importance of different income sources. For many of the early marchers, government transfer payments, such as Social Security, public assistance, and veterans benefits, were crucial. In the broad middle of the parade, households depended on wage or salary income from jobs or, less commonly, on entrepreneurial income from small businesses and professional practices. In the final moments of the procession, we saw marchers who are largely supported by their wealth in the form of income-producing assets such as stocks, bonds, and rental property.

5. *Occupation,* the marchers showed us, is a key determinant of household income, but not the overpowering factor we might have anticipated. From the reviewing stand, we saw low-skilled blue-collar, clerical, and service workers gradually give way to more skilled workers and then to managers and professionals. But there was considerable overlapping of occupational categories. We saw managers relatively early in the parade, and a few blue-collar workers marched with the Giants. One reason for this is that a two-income, blue-collar household can often outearn a low-level manager who does not have a working spouse. Another is that occupational pay scales overlap, even for very different occupations. For example, the top 25% of electricians earn more than the bottom 25% of aerospace engineers.

6. *Women's Shifting Role* was one of the defining features of the parade. At the beginning of the parade, we saw many older women and female heads of families. Among married-couple families some 20 minutes into the parade, wives without jobs were typical. But among the more prosperous households toward the end of

the parade, working wives were the rule. Only in the final moments of the parade did women's employment rates and the significance of their earnings decline.

7. *Minorities.* Blacks and Hispanics were at a disadvantage in the parade. They were overrepresented among the early marchers, often by female heads of households. On the other hand, given their traditional position in American society, their strong representation in the middle of the parade was probably surprising to many observers.

8. *Income and the Class Structure.* The parade suggests that the relationship between the distribution of income and the class structure is clear at the extremes but somewhat blurred in the middle. We can think about the problem in terms of the class model we introduced in Chapter 1. The people at the very beginning of the parade, who have no employment income or work at very low-wage jobs, correspond to our underclass and working poor. The towering marchers we saw in the last 2 minutes of the parade represent the top of the upper-middle class and the capitalist class. But in the middle of the procession, we found factory workers marching with middle-class managers. And we saw dual-income, working-class marchers looking down on single-income, upper–middle-class marchers. In sum, the class structure as we have defined it (or, for that matter, as Coleman and Rainwater defined it based on social prestige in Boston and Kansas City) does not exactly match the distribution of household income. The mismatch is greatest in the middle of the parade.

The Distribution of Income

Table 4.1, based on the annual Census Bureau survey, confirms our broad impression of the income parade: Almost 40% of households get along on fairly modest incomes (under \$35,000), 45% can be described as middle income (\$35,000 to

Table 4.1   Household Income, 2005

Income	All Households	Family Households
Under \$15,000	14.9	9.2
\$15,000–\$25,000	12.0	9.6
\$25,000–\$35,000	11.5	10.5
\$35,000–\$50,000	15.1	15.0
\$50,000–\$75,000	18.9	21.0
\$75,000–\$100,000	11.4	13.8
\$100,000–\$200,000	13.3	16.9
\$200,000 and over	3.0	3.9
Total	100.0	100.0
Number (in millions)	114.4	77.4

SOURCE: U.S. Census Bureau.

NOTE: Households include family households, individuals living alone, and unrelated persons sharing housing.

$100,000), the remaining households are relatively affluent (over $100,000), but incomes over $200,000 are rare. Even when we look separately at family households (excluding the typically less prosperous single-person households), the distribution is not radically altered, as the second column indicates.

Ethnicity and family structure create variants on this basic pattern. The income distributions of white, black, and Hispanic families are compared in Table 4.2. Note the high proportion of minority families at middle-income levels: around 40% of minority families fall between $35,000 and $100,000. The difference between these relatively comfortable middle-income families and the 50% of minority families struggling with lower incomes is often the difference between families headed by married couples and those headed by single females. This is particularly the case for black families.

Table 4.2   Family Income by Race and Hispanic Origin, 2005

Income	Percent		
	White	Black	Hispanic
Under $25,000	16.5	36.1	31.0
$25,000–$35,000	10.5	13.3	15.2
$35,000–$50,000	14.7	14.7	17.4
$50,000–$75,000	21.0	16.4	18.3
$75,000–$100,000	14.1	9.2	8.8
Over $100,000	23.2	10.3	9.3
Total	100	100	100
Median Income	$59,000	$35,000	$38,000
Number (in millions)	63.4	9.1	9.9

The bar graph (Figure 4.2) compares the median incomes of married couple- and female-headed families, by ethnicity. (The median income, as we noted earlier, is the midpoint in the distribution, dividing the top 50 from the bottom 50%.) The graph highlights the economic plight of female-headed families. It also reveals that the income gap between female- and couple-headed families is greater than that between majority and minority families. Of course, female-headed families are more prevalent among minority households—reflecting in part the economic strains to which they are subjected: 45% of black families, 18% of Hispanic families, and 15% of white families are female-headed. Nonetheless, most female-headed families are non-Hispanic white.

In this section and in the income parade, we have singled out female-headed families. What about male-headed families—that is, families headed by single men? They are less significant for our analysis because there are relatively few male-headed families (about 6% of all families), and their median income ($46,800) is close to that for married-couple families without working wives. On the other hand, one out of every six families is female-headed. The median income of such families ($30,650) is well below that of other family types.

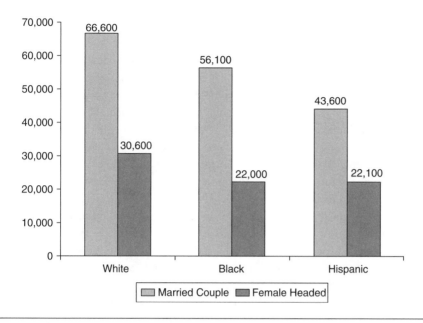

Figure 4.2   Median Family Income by Ethnicity and Family Type, 2005

Sources of Income

In the income parade, we noted shifts in the predominant sources of income. Table 4.3 traces this tendency. The story this table tells is simple but important. Wages and salaries provide most income for most people. But for the bottom 40% of households, a big chunk of income comes from government transfers such as social security, veterans benefits, and public assistance.[2] At successively higher levels, capitalist income (stock dividends, interest rents, and the like) and business profits provide increasing proportions of total income, until they exceed wage and salary income.

Table 4.3   Sources of Income, 2006

Income Group	Wage & Salary	Small Business	Capitalist	Govt. & Other	Total
Bottom Fifth	47	5	5	43	100
Second Fifth	61	3	4	31	100
Middle Fifth	73	3	4	21	100
Fourth Fifth	75	3	5	18	100
Top Fifth	60	11	15	14	100
Top 1%	35	20	33	11	100

SOURCE: Mishel et al., 2007:79. Based on Urban-Brookings Tax Policy Center estimates.

NOTE: *Small Business* = self-employment income, including business, profession, farm, partnerships, etc; *Capitalist* = interest, dividends, capital gains, rent, estate, and trust; *Govt. & Other* = mainly, government transfers including Social Security, public assistance, veterans benefits, etc.

---

[2]Middle- and upper-income households also receive government transfer payments, as the table suggests, but such payments form a modest proportion of their relatively larger incomes.

Income Shares

Aside from parades, the distribution of income is typically analyzed in one of two standard formats: (1) the distribution of households (or families) across ranges of income, and (2) the distribution of income shares among stratified segments of the population. Table 4.1 is a clear example of the first approach, which was the basic source for the income parade. The income shares approach could be called a slices-of-pie distribution. It conceives of the total income of all households as a national income pie, which has been sliced into pieces ranging from stingy to generous. This way of looking at income distribution measures the extent to which income is concentrated toward the top.

Figure 4.3 depicts the distribution of income shares among household income quintiles. The figure indicates the share of the total income pie that goes to each quintile. The first (poorest) quintile, for example, receives 4% of aggregate income. (Obviously, if the distribution of income were perfectly equal, each quintile would receive exactly 20%.)

Figure 4.3 reveals a remarkable concentration of income. The income share claimed by the richest fifth of households is 15 times that received by the poorest quintile. Such comparisons belie the common notion that the middle class receives the most income. Indeed, a modern Robin Hood could transform the lives of the poor and near-poor by transferring income from the richest 5% of households to the poorest 20% until he had equalized the incomes of the two groups. In 2005,

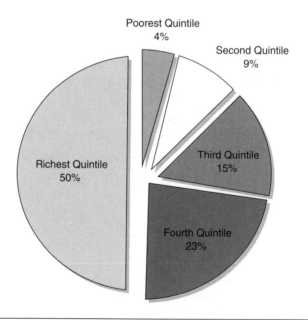

Figure 4.3   Shares of Aggregate Income Received by Quintiles of Households, 2000 (Pretax)

SOURCE: U.S. Census Bureau.

this tactic would have raised the average income of the bottom group from approximately $10,000 to $66,000.

Taxes and Transfers:
The Government as Robin Hood?

But isn't the government Robin Hood? Doesn't the government use progressive taxation to equalize incomes? The income distributions we have presented so far are all based on pretax data. What would they look like if we took taxes into consideration? The answer is, not terribly different.

We do have a federal personal income tax that is progressive in its overall effect—that is, people with higher incomes pay a greater proportion of their total income to the Internal Revenue Service than do those with lower incomes. The progressive tendency of the federal income tax is reinforced by the Earned Income Tax Credit, a generous provision of the tax code designed to help the working poor.

However, other taxes operate in the opposite direction. Chief among these regressive taxes are the sales taxes that are levied by states and localities. A sales tax imposes the same tax rate on a pair of children's shoes whether the purchasing parent is a low-wage worker or a millionaire. Because the low-wage worker spends a much higher proportion of her family's income on consumer items than does the millionaire (who is likely to reserve some income for investment), she loses a higher percentage of her income to the sales tax. The federal payroll deduction for Social Security is also regressive. For example, a worker earning $30,000 would pay more than 6% of his wages in payroll taxes. An executive earning $300,000 would pay less than 2%.

The combined effect of all federal taxes is relatively progressive, according to the estimates in Table 4.4. But, as the table indicates, posttax incomes remain steeply stratified. The effective tax rates listed in the table are not the familiar income tax brackets, but the proportion of income actually absorbed by federal taxes, including the income tax, payroll taxes, corporate taxes, excise taxes, and estate taxes. The remarkably low rate given in the table for the lowest-income households reflects the generous benefits low-income families dependent on low-wage workers receive from the Earned Income Tax Credit.

Aside from the redistributive effect of taxes, the government can play Robin Hood through transfer payments and noncash benefits. Because transfer payments, such as Social Security and public assistance, are counted as part of cash income, they are, unlike taxes, already reflected in the income data we saw in the last section. The value of noncash benefits, such as food stamps and Medicare (the federal health care program for the elderly), is not included in the income data.

The influence of transfer payments is generally progressive for the obvious reason that they often are specifically designed to help the poor and for the less obvious reason that a large share of transfer income goes to the elderly who tend to be at the lower end of the pretransfer income distribution. The same can be said of noncash benefits. In general, transfer payments and noncash benefits raise the living standard of the poorest households but do not dramatically alter the overall structure of economic inequality.

Table 4.4   Effective Federal Tax Rates on Households, 2004

| Households | Average Income | | Effective Federal Tax Rate |
	Pretax	Posttax	Percent
Bottom fifth	$15,400	$14,700	4.5
Second fifth	$36,300	$32,700	10.0
Middle fifth	$56,200	$48,400	13.9
Fourth fifth	$81,700	$67,600	17.2
Top fifth	$207,200	$155,200	25.1
Top 5%	$443,400	$317,500	28.5
Top 1%	$1,259,700	$867,800	31.1

SOURCE: Congressional Budget Office (CBO) 2006, Table 1.

So how effective is the government as Robin Hood? Figure 4.4 estimates the combined effects of taxes, transfers, and benefits on income shares. The "before" shares are based on pretax incomes stripped of transfer payments. The "after" shares were produced by adding in the value of cash transfers and noncash benefits and deducting federal and state income taxes, along with Social Security payroll taxes. In other words, we are looking at income inequality before and after the government Robin Hood has completed his work. The differences in income shares, as defined here, are modest in the broad middle, but notable for the top and bottom fifths. In particular, the share of the poorest fifth is four times higher than it would be without taxes and government programs. Despite its reduced share, the richest fifth still claims almost half of all personal income.

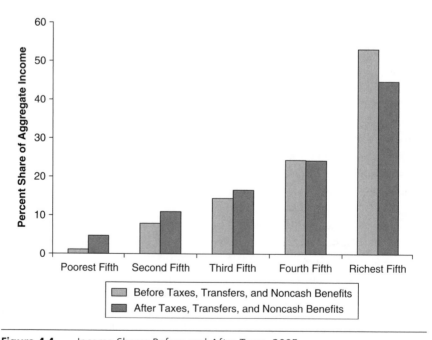

Figure 4.4   Income Shares Before and After Taxes, 2005

SOURCE: U.S. Census Bureau Web site. Table RDI-7, definitions 3 and 14.

How Many Poor?

Our discussion of income distribution and redistribution has skirted an important issue: how many people have such low incomes that they can be considered poor? The easiest answer is based on official government statistics, which recorded 37 million poor Americans in 2005, about 13% of the population. However, any count of the poor depends on the standard or definition of poverty used by the counters. Many researchers would adjust these figures upward or downward because they are skeptical of the standard employed by government statisticians. We take up the problem of defining poverty in Chapter 10, so that readers will be able to draw their own conclusions.

Women and the Distribution of Household Income

One lesson we drew from the income parade concerned the way women's situations changed with rising income. The elderly women we saw early in the parade were typically widows older than 75. Because women traditionally have had lower earnings and shorter, less continuous work histories, they have weaker personal claims on retirement income. Older women tend to be dependent on a husband's pension or Social Security check and can lose all or part of that income in the event of their divorce or his death. Since women tend to outlive men, they are more likely to survive long enough to use up their savings.

Although the more generous Social Security benefits of recent years have sharply reduced poverty rates among the elderly, older women living alone continue to have relatively high poverty rates. Only 4% of wives over 65 are poor, according to the Census Bureau. But among women over 65 and living alone, 21% are poor.

Largely as a result of elevated divorce rates and the growing proportion of children born to single mothers, nearly 18% of all U.S. families are now female-headed, compared with 10% in 1970. Among families with children, 22% are female-headed. The women who head these households face multiple disadvantages. Child care responsibilities make it difficult to work full time. Most single mothers and divorced or separated wives who are custodial parents receive meager child support payments or none at all. Generally, the economic situation of men improves after a marital separation, whereas that of women deteriorates (Hoffman 1977; U.S. Census Bureau 1994:33).

In 2003, according to the most recent U.S. Census Bureau (2006) report on custodial parents, (1) the average (mean) income of custodial mothers was $28,600, far below the average for all families; (2) the annual child support income received by custodial mothers averages about $3,600; (3) one-fourth of custodial mothers received no child support; and (5) 26% of custodial mothers and their children were surviving on incomes below the official poverty line.

Women who go to work to support their families often find themselves in the lower-paying pink-collar jobs described in Chapter 3. Although, as we noted there, women's earnings have advanced relative to men's, even women employed full time still lag well behind similarly employed men. Of course, responsibilities at home, especially for single mothers of young children, prevent many women from working full time, year round.

Working wives face the same problems in the labor market as female heads of families, and a similar clash between the roles of nurturer and breadwinner. Nonetheless, the percentage of married women in the labor force has been rising since the 1920s (Beeghley 1996:231). From 1960 to 2000, the labor force participation rate of wives doubled.

Most married women still earn less than their partners do, though the gap has been closing. More important, in an era when men's earnings are declining, the rising earnings of women are increasingly crucial to family incomes. Just how crucial is revealed in Table 4.5, which examines spouses' contributions to change in the incomes of families with children over two decades. Without the growth in women's earnings (often as a result of simply working longer hours), total income would have fallen between 1979 and 2000 for the bottom 40% of families and barely changed for the middle fifth. Only for the top fifth are the wives' increased earnings less than vital to the family's standard of living. (Women at this level have, in fact, increased their hours of work at least as much as women at other levels, but the relative impact of their increased earnings on an already high family income is modest.)

Table 4.5   Spouses' Contribution to Change in Family Income, 1979–2000

	Percent Change				
Income Growth	Bottom Fifth	Second Fifth	Third Fifth	Fourth Fifth	Top Fifth
Husbands' Contribution	−9.1	−1.1	5.8	16.2	54.1
Wives' Contribution	18.8	17.6	19.1	16.6	13.0
Total Change	9.1	16.6	24.9	32.8	67.1

SOURCE: Modified from Mishel et al. 2007:Table 1.25.

NOTE: Refers to married-couple families with children and head of household 25 to 54.

The Distribution of Wealth

We now turn from the distribution of income to the distribution of wealth. We have already distinguished these two concepts: Income is the inflow of money over a period of time. Wealth is the value of assets held at a point in time. One year's wages, interest, and dividends, such as might be reported on a federal income tax return, are examples of income. The real estate, bank account, and stock shares someone owned on, say, December 31, 2009, are examples of wealth.

Wealth, we might say, is nothing more than accumulated income. True, but this underestimates the significance of wealth as a distinct dimension of class inequality. Wealth, unlike income, need not be consumed in paying for daily necessities. It enhances what Max Weber called "life chances" in more basic ways. Wealth offers safety net protection against a sudden drop in living standard in the event of job loss or other emergency. Most families do not hold significant wealth and would be in desperate straits if they missed 1 or 2 months' paychecks. Wealth can be converted into home ownership, business ownership, or college education. As we saw earlier in the chapter, people with very high incomes typically derive most of their income from wealth in the form of stock dividends, bond interest, commercial real estate rents, and other capitalist sources.

Wealth provides an important mechanism for the intergenerational transmission of inequality. As we will see in Chapter 8, about half of the wealthiest people in America inherited family fortunes. On a more modest level, the high school student who knows that there is money in the bank to pay for her college education and the young couple who purchase a house with help from their parents are the beneficiaries of the wealth accumulated by previous generations. Most Americans can expect, at best, a modest inheritance. From this perspective, it is hardly surprising that upwardly mobile African Americans with comfortable incomes lag far behind white peers in wealth (Oliver and Shapiro 1995).

Wealth is measured in two ways: gross assets and net worth. The concept of gross assets refers to the total value of the assets someone owns. Net worth, a more realistic concept, is the value of assets owned minus the amount of debt owed. The net worth of most households, according to the Federal Reserve survey summarized in Table 4.6, is relatively modest. Even including the value of home equity, fewer than half of households have net worth greater than $100,000. Most households have little in the way of capitalist assets, such as stocks, bonds, or commercial real estate. They derive the greater part of their net worth from three asset types: home equity, car equity, and bank deposits (Kennickell 2006; Mishel et al. 2007:Ch. 5; U.S. Census 2003).

We can distinguish three broad classes of wealth holders:[3]

1. *The Nearly Propertyless Class.* About 40% of households, with net worths under $50,000 in 2004. Most are worth less than $25,000, and some have a negative

---

[3]This discussion draws on Bucks, Kennickell, and Moore 2006; Kennickell 2006; Mishel et al. 2007; U.S. Census 2003; and Wolff 1998. The figures are for 2004 and are given in 2004 dollars.

Table 4.6   Households by Net Worth, 2004

	Percent of Households
Under $5,000	18.3
$5,000–$25,000	12.2
$25,000–$50,000	8.8
$50,000–$100,000	11.9
$100,000–$250,000	18.6
$250,000–$500,000	12.4
$500,000–$1 million	9.6
$1 million and above	8.1
Total	100.0

SOURCE: Kennickell 2006:8.

net worth because they owe more than they own. The great majority have automobiles and about 60% own their homes. But this is a debt-ridden class: what they owe is high relative to the value of their assets. Younger families and most African American and Hispanic households fall into this wealth class.

2. *The "Nest Egg" Class.* About 50% of households, with net worths ranging from $50,000 to $800,000. The families in this class might be described as savers rather than investors. Their debt is modest. They own safe, interest-earning assets such as savings accounts, CDs, and government bonds. They accumulate substantial retirement savings in IRA and 401k accounts, but their largest single asset is likely to be the home they live in.

3. *The Investor Class.* Just 10% of households, worth more than $800,000. The households in this class own most of the privately held investment assets and typically control substantial portfolios of stocks, mutual funds, and bonds. Many members of this class have interests in small businesses, professional practices, and commercial real estate. On the other hand, equity in homes and autos contributes a modest proportion to their net worth. The investor class is relatively free of debt. Although it controls most of the total of gross assets owned by households, it is responsible for a very small proportion of total liabilities.

As the privileged finances of our top class suggest, wealth is highly concentrated—much more concentrated than income. For example, in 2004 the top 1% of income earners received 17% of aggregate income, while the top 1% of wealth holders owned about 34% of net worth (Mishel et al. 2007:251). The concentration of wealth at the top is so great that the top 1% now holds more net worth than the bottom 90% (Table 4.7).

Another basic conclusion that can be drawn from studies of wealth is that investment assets are much more concentrated than consumption-oriented assets such as automobiles and owner-occupied homes (Table 4.8). Ownership of corporate stock

Table 4.7   Concentration of Wealth, 2004

	Share in Percent	
	Net Worth	Net Financial Assets
Top 1%	34.3	42.2
Next 9%	36.9	38.7
Bottom 90%	28.7	19.1
Total 100%	100.0*	100.0

SOURCE: Mishel et al. 2007:251, based on analysis by Edward Wolff of the Survey of Consumer Finances.

*Total may not add up due to rounding.

and mutual fund shares, bonds, investment real estate, and small business equity is almost entirely concentrated in the hands of the top 10% of wealth holders. The value of home and auto equity is much more broadly distributed. In Chapter 8, we consider some political implications of this remarkable concentration of economic power.

Table 4.8   Concentration of Key Assets, 2004

		Share Held	
	Top 1%	by Top 10%	Bottom 90%
Widely Held Assets			
Home Equity	12.6	45.9	54.1
Vehicles	5.8	23.3	76.6
Bank Accounts	29.5	63.1	36.8
CDs	11.5	55.3	44.6
Concentrated Assets			
Stocks & Mutual Funds	36.8	78.8	21.2
Bonds	64.1	91.7	8.3
Investment Real Estate Equity	47.3	83.0	17.0
Business Equity	62.3	90.4	9.5
Net Worth	**33.4**	**69.5**	**30.4**

SOURCE: Based on Kennickell 2006, Table 11a.

Trends in the Distribution of Wealth

Sometime in the early 1970s, a great shift began in the distributions of wealth and income, paralleling the growing disparities in job earnings we examined in the last chapter. We recognized this transformation in the distinction we made earlier between the post–World War II Age of Shared Prosperity and the current Age of Growing Inequality.

The trend toward increasing inequality in the distribution of wealth was especially notable in the 1980s. At the top of the wealth pyramid, the combined net worth of the people on *Forbes* magazine's list of the 400 richest Americans more than doubled in real dollar value in the 1980s and again in the 1990s. By 2006, the poorest person on the *Forbes* list was worth a billion dollars, and the richest (Microsoft founder Bill Gates) was worth $53 billion. Their combined net worth was then an awe-inspiring $1.259 billion, a figure on the order of magnitude of the federal budget (*Forbes* 1996, 2001a, 2006).

On a broader scale, the wealth of the richest 1% (roughly 1 million households) has grown spectacularly since the mid-1970s. Figure 4.5, which we previewed in Chapter 1, traces the proportion of aggregate net worth held by the top 1% since the 1960s. The curve assumes the familiar U-shape trajectory, bottoming out in the 1970s and then climbing steeply in the 1980s. By the late 1990s, the top 1% held almost 40% of net worth—more, as we have seen, than the bottom 90% of households, and probably more than at any time since the 1930s (Wolff 1993, 1998).

At the same time that the wealthy have become wealthier, more Americans have become wealthy. The number of millionaire households (in inflation-adjusted dollars) more than doubled in the decade following 1995. By 2004, 9 million American families were worth more than $1 million and 110,000 households were worth more than $25 million (Frank 2007).

What accounts for these remarkable increases in the number of wealthy families and the concentration of wealth? Four factors seem especially important: (1) the rapid growth of incomes at the top of the distribution, which has enabled those with high incomes to accumulate new funds for investment while incomes at lower levels have stagnated; (2) more specifically, the emergence of new fortunes associated with information technology and other high-growth sectors of the economy; (3) declining tax rates for the top income bracket, allowing high-income households to retain more of what they make; and (4) the generally rising stock market since the early 1980s, which has swelled the value of securities held by the wealthy.

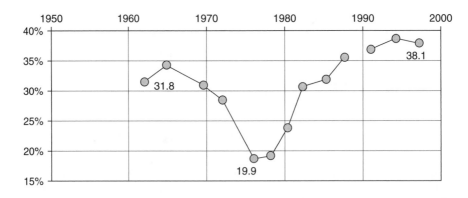

Figure 4.5   Share of Net Worth Held by Top 1% of Households

SOURCE: Wolff 1993, 1998.

Trends in the Distribution of Income

During the Age of Shared Prosperity, family incomes at all levels were growing at a brisk pace and gradually becoming more equal. After the early seventies, income growth tapered off, and the fortunes of American families began to diverge.

The transition from the Age of Shared Prosperity to the contemporary Age of Growing Inequality is reflected in Figure 4.6, another of the curves we previewed in Chapter 1. This graph uses income ratios to measure the advantage of the richest 5% of families over the bottom 40%. For example, in 1950, the average income of the top 5% (in the valuable dollars of that era) was $13,200, which is more than 8 times the $1,600 average income of the bottom 40% in 1950. The graph shows that this ratio between the top 5 and the bottom 40 dipped in the 1950s and 1960s when low-income families were gaining on the rich; it climbed after the mid-1970s as income growth accelerated at the top and low-income households were left behind. By the end of the century, the ratio between the incomes of the bottom 40% and the top 5% was the highest ever measured in 50 years of government income surveys.

Figure 4.7 tells us what the 1970s income shift meant for people at different levels of income. The bars represent the percentage increase in real incomes at each level during the Age of Shared Prosperity and the Age of Growing Inequality. Comparing the side-by-side panels, we can see a reversal of the image. Two conclusions are obvious: (1) income growth was generally higher and more broadly shared in the first period, and (2) growth was fastest at the bottom in the first period and fastest at the top in the second period.

Imagine the shifting fortunes of three families under the conditions described by Figure 4.7. The first family, in the poorest fifth, sees its income more than double (120% growth) in the first period, but practically stagnate in the second. The next family, in the middle fifth, undergoes a similar but less dramatic shift from high

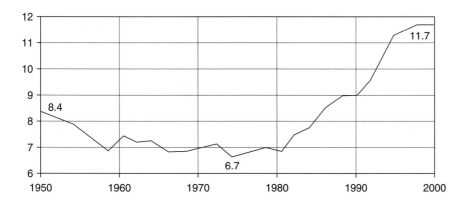

Figure 4.6   Income Ratio: Top 5% Versus Bottom 40% of Families

SOURCE: U.S. Census Bureau.

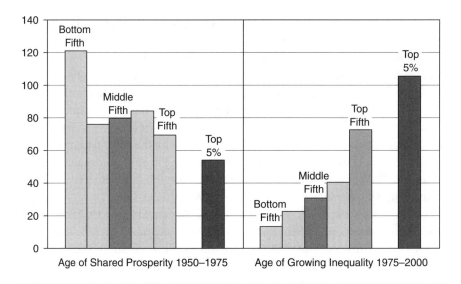

Figure 4.7   Change in Family Income for Fifths and Top 5% of Families,
1950–2000 (Percent Growth)

SOURCE: U.S. Census Bureau.

growth to slow growth. The third family, in the privileged top 5%, finds moderate
gains in the first period and a doubling of income in the second.

Although income growth in recent years has been slower and more concentrated, growth at the top has allowed an expanding proportion of families to attain
relative affluence. As Figure 4.8 shows, the proportion of households with incomes
greater than $100,000 has grown steadily since the 1960s.

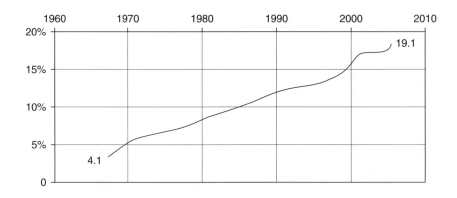

Figure 4.8   Households With Incomes Over $100,000

Source: U.S. Census Bureau.

Income Dynamics

The three hypothetical families whose fortunes we traced from 1950 to 2000 had one thing in common: their relative positions in the income distribution were stable, even when their incomes were changing. We assumed, for example, that the family that started in the bottom quintile was still there several decades later. We almost automatically make this kind of assumption when we talk about shifts in the distribution of income. But the government income surveys we have been analyzing in this chapter do not follow families over time. They are, in effect, periodic snapshots of the income distribution, which tell us nothing about the degree to which individual families are moving up or down relative to one another.

Following the incomes of specific families over time is difficult and expensive. The few studies that have done so reveal a surprising amount of movement. From one year to the next, a family's income may change abruptly because someone lost a job or a spouse rejoined the labor force. Over longer periods, earnings tend to expand with experience and successive promotions. For example, during an academic career, the salary of a college professor might double in real-dollar terms. Even the earnings of low-skilled workers tend to rise over time, though more slowly than those of professionals.

Table 4.9, based on a study that has followed several thousand families over three decades, shows considerable movement ("income mobility") in the distribution of family income. In the 1990s, for example, approximately 30% of families moved to a higher income quintile and about the same proportion moved down. But the table also reveals a slowing of income mobility: fewer families are moving up or down; an increasing proportion end the decade where they began. Closer examination of the data from all three decades shows that most movement, in either direction, was short range, from one quintile to an adjacent quintile. Only rarely do families rocket from the bottom quintile to the top or plunge from the top to the bottom in the course of a decade.

Table 4.9   Family Income Mobility

	Change in Income Quintile, in Percent			
	Up	*Stable*	*Down*	*Total*
1970s	33.0	35.7	31.3	100.0
1980s	32.4	37.0	30.6	100.0
1990s	30.1	40.4	29.4	100.0

SOURCE: Author's analysis of Table 2.3, Mishel et al. 2007:106.

Changing Federal Tax Rates

The trend toward greater income inequality since the 1970s was magnified by regressive changes in the federal tax system. (Census Bureau income figures are, as noted earlier, pretax.) The top tax rate on personal income plunged from 77 to as little as 28%. Corporate and inheritance taxes, whose main effects are felt by the wealthy, were also reduced. At the same time, payroll taxes, paid largely by lower- and middle-income workers, were jacked up (U.S. House of Representatives 1991a). The one change that ran against the generally regressive tide was the Earned Income Tax Credit, described earlier, designed to help the working poor. The result of these policies can be seen in Table 4.10, which traces the evolution of effective tax rates— that is, the proportion of income lost to the combined effect of all federal taxes.

The table misses the effects of significant tax reductions implemented before 1979 but allows us to draw some conclusions about subsequent trends. Over the period taken as a whole, effective tax rates have come down for average taxpayers at all levels. But the most significant changes came at the very top. Effective tax rates for the richest 1% have moved up and down as conservative Republican and liberal Democratic administrations have shifted the tax code back and forth. But over the whole period, households at this level have received the biggest tax relief in both relative and absolute terms. The average 5.9% effective tax reduction for the average household in the top 1% would mean a tax savings of $62,300 in 2004. Corresponding savings for the middle fifth and bottom fifth of families amount to just $2,500 and $530, respectively.[4]

Wealthy households could expect further relief when the provisions of the 2001 tax legislation signed by President George W. Bush are fully phased in. The most dramatic change would be the gradual and finally total elimination of inheritance (estate) taxes by 2010. Since only estates in excess of $1 million have been taxed in recent years, this change will be meaningless for all but 1 or 2% of Americans. But the savings will be huge for the Mars family, owners of the Mars candy company. In 2006, the combined net worth of the three wealthiest family members was $31.5 billion, about half of which would have gone to estate taxes. As we will see in Chapter 8, the Mars family played an important role in the lobbying campaign to eliminate the tax. (The Mars family may feel compelled to continue its efforts

Table 4.10   Effective Federal Tax Rates for Families, 1979–2004

	1979	1985	1991	1997	2004	Change 1979–2004
Bottom Fifth	8.0	9.8	8.4	5.8	4.5	−3.5
Second Fifth	14.3	14.8	14.2	13.6	10.0	−4.3
Middle Fifth	18.6	18.1	17.6	17.4	13.9	−4.7
Fourth Fifth	21.2	20.4	20.5	20.5	17.2	−4.0
Top Fifth	27.5	24.0	25.3	28.0	25.1	−2.4
Top 5%	31.8	25.4	27.6	31.6	28.5	−3.3
Top 1%	37.0	27.0	29.9	34.9	31.1	−5.9

SOURCE: CBO 2006.

---

[4]Calculated from average incomes for each group given in CBO 2006.

because new legislation will be required to carry estate tax relief beyond 2010. Without it, rates will revert to previous levels.)

Conclusion

In this chapter, we added evidence of growing inequality in the distribution of income and wealth since the early 1970s to the evidence of growing inequality in earnings that we explored in the last chapter. The polarization of incomes is all the more remarkable because it reverses a well-documented trend toward greater income equality from the 1930s into the 1970s (Miller 1971; U.S. Census Bureau 1996b). How can we account for the shift? This question is an enlarged version of the one we asked at the end of Chapter 3. There, we were interested in the increasing disparity in earnings. Here, we were concerned with all sources of income and with whole households rather than individual workers. In this section, we review what we have learned in this chapter, giving particular attention to developments that can help explain change.

We began the chapter with an imaginary income parade, a device to visualize the income distribution and the factors that shape it. The parade began with a long line of small people—not just the poor but also millions of families living marginally on the earnings of low-wage workers. At the end of the parade, we were struck by the abrupt increase in the size of the marchers, who grew in a matter of minutes to astronomical proportions, reflective of astronomical income.

The changing mix of occupations during the course of the parade was about what we expected. More surprising were the nonoccupational factors that powerfully influenced where people marched in the ranks—in particular, the sources of household income and the number of workers in a family. Most households, of course, depend on job earnings. We noticed that single-worker families were common in the first half of the parade. Much later in the procession, among people with incomes around $100,000, we found very few families without two employed adults. But we discovered that jobs were less significant for those at the beginning and the very end of the procession. The first marchers were heavily dependent on government transfers, from Social Security to public assistance. The very last marchers—especially those with incomes above $1 million—depended less on jobs than on investments for their incomes.

If those at the end of the parade are most dependent on financial wealth for their incomes, the rising concentration of wealth is certainly strengthening income inequality. Of course, the accumulation of wealth is also a *result* of income inequality—as well as the changes in effective tax rates that enabled those with the highest incomes to retain a higher proportion of their incomes.

The parade focused attention on the social factors leading to increased income inequality. We noted, for example, that families headed by females were crowded into the early part of the parade. The prevalence of such families is growing as a result of increased divorce and births to single mothers, contributing inevitably to the growth in income inequality. These social trends are strengthening the economic pressures toward polarization that we discussed in Chapter 3. The result is what we have called an Age of Growing Inequality, which we describe more broadly in chapters that follow.

Key Terms Defined in the Glossary

Age of Growing Inequality   mean/median
Age of Shared Prosperity   nearly propertyless class
distribution of income   (see wealth classes)
distribution of wealth   nest-egg class (see wealth classes)
dividend   net worth (see wealth)
Earned Income Tax Credit (EITC)   progressive tax
earnings   quintile
effective tax rates   real income
government transfers   regressive tax
gross assets   wealth
income   working rich
investor class (see wealth classes)

Suggested Readings

Collins Chuck and Felice Yeskel, with United for a Fair Economy and Class Action. 2005. *Economic Apartheid in America: A Primer on Economic Inequality and Insecurity.* Revised ed. New York: New Press.
> *A well-documented primer on economic inequality in the United States, its consequences, and political cures.*

Fine, Michelle and Lois Weis. 1998. *The Unknown City: The Lives of Poor and Working-Class Young Adults.* Boston: Beacon.
> *Family and economic lives of white, black, and Hispanic young adults in the context of two economically troubled cities in the Northeast. Most are working poor or underclass.*

Frank, Robert. 2007. *Richistan: A Journey Through the American Wealth Boom and the Lives of the New Rich.* New York: Crown.
> *Despite the hokey title, a well-informed examination of the fortunes, spending habits, social lives, and politics of the new rich by a reporter who has covered them for the* Wall Street Journal.

Mishel, Lawrence, Jared Bernstein, and Sylvia Allegretto. 2007. *The State of Working America 2006/2007.* Ithaca, NY: Cornell University Press.
> *Long-term trends in the distribution of income and wealth, and other topics. Updated editions published regularly.*

Rose, Stephen. 2007. *Social Stratification in the United States: The American Profile Poster.* Revised and updated ed. New York: New Press.
> *Ingenious poster with companion booklet, illustrating the distribution of income, occupation, and household types.*

Shapiro, Thomas. 2004. *The Hidden Cost of Being African American: How Wealth Perpetuates Inequality.* New York: Oxford University Press.
> *The role of wealth and inheritance in perpetuating racial inequality.*

U.S. Bureau of the Census. (n.d.). *Money Income of Households, Families, and Persons in the United States.* Current Population Reports. Consumer Income. Series P-60.
> *Published annually, the reports in this series are the basic source of detailed information on the distribution of income. (The title varies.) Text available at www.census.gov.*

CHAPTER 5

Socialization, Association, Lifestyles, and Values

Let me tell you about the rich. They are different from you and me.

F. Scott Fitzgerald

Yes, they have more money.

Ernest Hemingway

T
he title of this chapter might have been "Does Class Matter?" So far, we have examined the distribution of income and wealth, the occupational structure, and conceptions of the class structure. But, aside from a hint here or there, we have not shown that class really matters in our everyday lives. In this chapter, we demonstrate that class shapes our experience from early childhood to mature adulthood. In fact, we will find that class position affects some of the most intimate aspects of our lives.

The discussion follows up on Max Weber's ideas about status communities, Lloyd Warner's description of social life in Yankee City, and the Lynds' account of the emergence of the wealthy X clan in Middletown in earlier chapters. These authors noted that people with similar class positions tend to draw together. They live in the same neighborhoods, develop friendships, spend leisure time together, and join the same clubs and churches. Their children go to the same schools, form social cliques, become teammates, form romantic attachments, and grow up to marry one another. Gradually, shared experiences become the basis for a distinctive lifestyle and common set of values, which parents pass on to their young. They develop, in other words, a self-perpetuating class subculture.

Our discussion emphasizes two of the basic variables we mentioned in Chapter 1: *socialization,* the learning process that prepares new members of society for social life; and *association,* social patterning of interpersonal contact.

Bourdieu: The Varieties of Capital

The work of French sociologist Pierre Bourdieu (1986) on class reproduction and the varieties of capital will help us understand the larger significance of the material covered in this chapter. Capital may be defined as value accumulated over time and capable of yielding future benefits. We are accustomed to thinking of capital as another name for wealth. But Bourdieu extends the concept. He distinguishes three forms of capital: *economic capital,* the basic monetary form, institutionalized as property rights; *cultural capital,* knowledge in its broadest sense, institutionalized as educational credentials, but encompassing such matters as table manners and how to swing a tennis racket; and *social capital,* mutual obligations embodied in social networks such as kinship, friendship, and group membership.

Bourdieu emphasizes that the value of each of these forms of capital is enhanced by its capacity for transformation into one of the others. For example, before he entered politics, George W. Bush took advantage of his social capital—the extensive social connections he had developed growing up in a prominent upper-class family— to gather economic capital for a series of business ventures (Kelly 2004). Generations of novelists have entertained readers with characters who strive to develop cultural and social capital to further their social ambitions.

From Bourdieu's perspective, the sum of the various forms of capital is the cumulative advantage of the privileged classes. It is a key to the reproduction of the class system by transmission from generation to generation. As several of the authors we examine in this chapter confirm, the class advantages or disadvantages that a child inherits are not just economic, but also, as Bourdieu would have it, cultural and social.

Children's Conception of Social Class

As they grow up, children absorb from their elders increasingly sophisticated notions about social class. An early study of primary school students in a New England town of 15,000 showed that by the sixth grade, children understood the class significance of items such as an English riding habit, an elegantly furnished room, tattered clothing, and different occupational activities, all presented to them in pictures. And when they were asked to place their peers in one of three classes, the sixth graders agreed 70% of the time with adults who rated parents from the same households (Stendler 1949).

Simmons and Rosenberg (1971) demonstrated that young children have a clear conception of occupational prestige differences. Even the third graders in their sample from the Baltimore city schools ordered 15 occupations from the National Opinion Research Center (NORC) list in a way that correlated almost perfectly with the rankings in a national survey of adults.

Subsequent studies have concentrated on the development of conceptions of class distinctions as children grow up. Tutor (1991) showed first, fourth, and sixth graders photographs of upper-, middle-, and lower-class people. She asked the children to group the adults and children depicted into families and match them with the corresponding pictures of cars and houses. The first graders did substantially better than chance at this task, the sixth graders produced near-perfect scores, and the fourth graders were not far behind.

Leahy (1981, 1983) probed the developing conceptions of "poor people" and "rich people" held by children ages 7 to 17. He found that young children conceive of the rich and the poor in overt, physical terms, while older children think in terms of psychological characteristics of individuals and their positions in the society. Here are some of their observations.

Joe, age 6:

> [Poor people have] no food. They won't have no Thanksgiving. They don't have nothing. . . . [R]ich people have crazy outfits and poor people have no outfits.

Mary, age 6:

> [People can become rich by going] to the store and they give you money. . . . [Or] if your husband gives you money, and your grandmother or your grandfather.

Pete, age 10:

> [People are rich] because they save their money and they earn it. They work as hard as they can and don't just go around and buy whatever they want.

Dean, age 12:

> I think that [rich and poor people] should all be the same, each have the same amount of money because then the rich people won't think they are so big.

In general, these studies show that even preschool children are aware of class differences. They suggest that as children grow older, their ideas about stratification become more consistent, abstract, and "accurate." By the time they reach 12 years of age, children are not very different from adults in their thinking about class.

Kohn: Class and Socialization

While studies such as those just reviewed approach socialization through the child's developing conception of the social world, a separate research tradition focuses on the child-raising practices of parents. For decades, studies of the latter type have recorded class differences in the way people raise their children (Bronfenbrenner 1966; Gecas 1979). Annette Lareau, whose research we examine in the next section, and Melvin Kohn have made intriguing contributions to this literature.

Kohn studied class differences in the values parents impart to their children. He wanted to understand exactly why such differences exist and how they contribute to the perpetuation of the class system (Kohn 1969, 1976, 1977; Kohn and Schooler 1983). In a series of surveys in the United States and abroad, Kohn and his associates asked parents to select from a list of characteristics those they considered most desirable for a child of the same age and sex as their own child.

Here are a few examples (Kohn 1969:218):

That he is a good student.

That he is popular with other children.

That he has good manners.

That he is curious about things.

That he is happy.

Although the studies found consensus across class levels about the importance of some values (parents of all classes wanted their children to be happy), there was less agreement about others. For example, parents at higher class levels were more likely to choose "curiosity," and those at lower class levels were more likely to select "obedience." The top panel (A) of Table 5.1 compares values characteristically cited by parents in the upper and lower halves of the class structure (for convenience labeled middle class and working class). Parental views were quite varied at every class level, but the values we are calling working class become increasingly common at lower class levels and those we have labeled middle class become more common at successively higher class levels.

Table 5.1   Typical Class Patterns in Parental Values and Occupational Experience

Middle-Class Pattern	Working-Class Pattern
A. Parents' Values for Children	
Self-Control	Obedience
Consideration of Others	Manners
Curiosity	"Good Student"
Happiness	Neatness, Cleanliness
B. Parents' Own Value Orientations	
Tolerance of Nonconformity	Strong Punishment of Deviant Behavior
Open to Innovation	Stuck to Old Ways
People Basically Good	People Not Trustworthy
Value Self-Direction	Believe in Strict Leadership
C. Job Characteristics	
Work Independently	Close Supervision
Varied Tasks	Repetitive Work
Work With People or Data	Work With Things

Families were assigned to classes on the basis of father's occupation. The researchers found that mothers' value preferences for children reflected their husbands' occupations. But they also observed that the class patterning of parental values could be strengthened or diluted according to the occupations of employed mothers. For example, among women married to working-class men, mothers with manual jobs were much more likely to conform to the working-class value pattern than were mothers with white-collar jobs.

Kohn interpreted the class patterns of parental values for children as follows: The middle-class parents who stress the values of self-control, curiosity, and consideration are cultivating capacities for self-direction and empathetic understanding of others in their children. The working-class parents who focus on obedience, neatness, and good manners are instilling behavioral conformity. The middle-class pattern—particularly in the emphasis laid on happiness, curiosity, and consideration—is oriented toward the *internal* dynamics of the person, both the child and others. The working-class pattern, on the other hand, assumes fixed *external* standards of behavior. This general difference is neatly illustrated in the top panel (A) of Table 5.1 by three pairs of contrasting values, beginning with "self-control" and "obedience." In each case, the first (middle class) choice favors internal development, and the second (working class) emphasizes conformity to external rules.

An additional finding substantiates this interpretation. Parents were asked about the specific sorts of misbehavior for which they would discipline their children. Their responses revealed that middle-class parents were more likely to punish a child for the *intent* of his or her behavior, in contrast with working-class mothers, who were more likely to discipline for the *consequences* of behavior. For example, a middle-class mother might penalize her child for throwing a temper tantrum, while a working-class mother penalizes for boisterous play. The first suggests a loss of internal control, the second a violation of external standards.

Kohn labeled the two underlying patterns *self-direction* and *conformity*. At successively higher class levels, he concluded, parents value self-direction more and conformity to external standards less. But what are the roots of these class differences? Kohn hypothesized that they reflect generalized value orientations that develop out of a specific aspect of social class: occupational experience. He reasoned, for example, that people who hold professional and managerial jobs, which are relatively unsupervised and require considerable exercise of individual judgment and initiative, are more likely to value self-direction than those who work at highly routinized, blue-collar jobs. In brief, self-direction at work should produce self-direction in values.

Evidence from the surveys supported these ideas. They showed, for example, that parents' general judgments about authority, deviance, and the goodness of human nature are related to social class (Table 5.1, Panel B). In particular, Kohn noted that "authoritarian attitudes" stressing "conformance to the dictates of authority and intolerance of nonconformity" become more frequent at lower class levels (Kohn 1969:79).

Finally, the surveys demonstrated that these general attitudes are systematically related to the character of respondents' occupational experience. Men whose work is (1) closely supervised, (2) repetitive, and (3) oriented toward things rather than people or data are the most likely to subscribe to authoritarian values and to judge jobs on their extrinsic qualities. Of course, what are ordinarily considered working-class jobs are most likely to fit this occupational pattern—though some (plumber) do not fit it as well as certain menial office jobs (data entry operator). Kohn also notes that the wives of men who share this occupational experience tend to hold similar values.

The results and interpretative logic of Kohn's research are summarized in Table 5.1. Remember that the middle-class side of this chart represents patterns that are increasingly frequent at higher class levels, and the working-class side describes patterns that become more frequent at lower class levels. (We are dealing with statistical tendencies here, not absolute contrasts between classes.) Reading the table from the bottom up on the middle-class side, we find that parents at higher class levels are more likely to work at jobs requiring intellectual flexibility and independent judgment (Panel C), more open to innovation and tolerant of nonconformity (Panel B), and more likely to encourage self-direction in their children (Panel A). The parallel finding on the working-class side is that parents at lower class levels are more subject to authority and routinization at work, more authoritarian in their judgments, and more likely to favor conformity in their children. These two contrasting patterns fit the causal chain that Kohn had anticipated to explain the relationship between social class and socialization patterns: Occupational experience gives rise to general value orientations, which in turn shape parental value preferences for children.

Further scrutiny of the data revealed that a second aspect of social class, level of education, exercises an independent influence on parental value orientations and value preferences for children and thus reinforces the class patterning of socialization. Kohn observed that education appears to "provide the intellectual flexibility and breadth of perspective that are essential for self-directed values" (1969:186). Kohn found that education and occupational conditions have independent impacts on parental values, although the effect of occupational conditions is substantially stronger.

Kohn's research on class differences in the socialization of children has important implications for our understanding of the class system as a whole. Since Marx, sociologists have been aware that life experience, especially occupational experience, shapes social values. Kohn (1969) observed that "the essence of higher class position is the expectation that one's decisions and actions can be consequential; the essence of lower class position is the belief that one is at the mercy of forces and people beyond one's control, often, beyond one's understanding" (p. 189). If this is true, we should expect people in top positions to learn to value self-direction and those at the bottom to learn to value conformity to authority. We might also anticipate that they will teach these values to their children.

At this point, the larger significance of Kohn's work becomes clear. When parents inculcate values that reflect their experience of the class system, they are preparing their children to assume a class position similar to their own and, by so doing, are contributing to the long-term maintenance of the class system. Kohn explicitly rejects the notion that these outcomes reflect the conscious intentions of parents. Working-class parents may, in fact, believe that the values they are encouraging in their children will help them to ascend the class hierarchy.

Lareau: Child Rearing Observed

Like Kohn, Lareau (2003) was interested in class differences in socialization. But Lareau went one step beyond Kohn. In addition to interviewing parents, she and a team of research assistants spent hundreds of hours observing parents and their 9- or 10 year-old children—usually at home, but also during routine activities outside the home. Such "naturalistic observation" is rare because it is expensive and time-consuming. It has the methodological disadvantage of generalization from an inevitably small sample (the team interviewed 88 families and observed 12). But observation can also be enormously rewarding, as it was for Lareau, because it yields information unfiltered by respondents and exposes researchers to important aspects of social life they may not have thought to ask about. Lareau's team asked families they observed not to treat them as guests, but to carry on their normal daily lives. They hoped to be as inconspicuous and taken-for-granted "as the family dog," and they seem, by and large, to have succeeded.

The study focused on families at three class levels, which Lareau labels middle class, working class, and poor. Her sample includes both black and white families at all three class levels. Judging from the high incomes and managerial or professional jobs she reports for the first group of families, Lareau's middle class could better be described as *upper*-middle class. Her working-class families seem to be a mix of what we would call working class and working poor. There is, then, a considerable economic gap separating Lareau's top class and the other two classes in this study. Perhaps it is not surprising that Lareau finds a corresponding gap in child-rearing practices.

Generalizing from rich observational data, Lareau describes two basic approaches to child rearing, which we can label *cultivated growth* and *natural growth*.[1] Parents who take the first approach hover over their children, scheduling their activities, fostering their talents, reasoning with them, and intervening on their behalf. Parents who take the second want to provide a safe and stable environment within which they expect the child to develop naturally; they guide their children with clear directives but allow them considerable autonomy in their everyday activities. Lareau reports that the upper–middle class families she observed practice cultivated growth, while both the working class and poor families practice the natural growth pattern of child rearing.

Lareau's observations cluster around three facets of children's lives: the organization of daily activities, the use of language, and relations with institutions such as schools.

Daily Activities. The upper–middle class 9- and 10-year-olds Lareau studied spend much of their time with adults or in activities organized by adults. The monthly calendar on the refrigerator door records a hectically scheduled life, with times for soccer practice, piano lessons, swim team, church choir, school play rehearsal, Girl Scouts, gymnastics, and violin lessons. There are scheduled play dates. One boy complains, "My mother signs me up for everything!" but says his activities make him feel "special" and admits he would be "bored" without them. On their own, these upper–middle class children are not sure how to use the limited free time they have.

Working-class and poor children, Lareau finds, lead slower, less structured lives. Much of their time is their own, and unlike their upper–middle class peers, they have no trouble entertaining themselves, generally in informal play with neighborhood children and cousins. Their parents do not, by and large, involve them in organized activities. Often, parents lack the prerequisite resources of time, money, and transportation. But many do not see the value of such activities, which upper–middle class parents regard as character building. Lareau finds that working class and poor parents regard the child's world and the adult world as distinct realms. Other than providing for their children's safety, they take only limited interest in the former.

Language. Lareau's interest in language extends more generally to the way that parents interact with children. Upper–middle class parents, she finds, engage in almost continual conversation with their children when they are together. They are intent on cultivating their child's facility with language. They teach children to express their own views and to believe that their opinions matter. They encourage

---

[1]We have substituted these simplified labels for Lareau's more cumbersome "concerted cultivation" and "accomplishment of natural growth."

them to ask questions of other adults and people in authority like teachers and doctors. In upper–middle class families, language is the main mechanism of discipline. Parents reason with children. Even when issuing directives, they attach reasons to them. And children learn to negotiate with their parents for what they want. (When negotiation is, from the children's viewpoint, unsuccessful, they often resort to whining, a tendency that the researchers did not observe with children at lower class levels.)

Conversation between parents and children in the working-class and poor homes was much less extensive. These parents and children were often silent in each others company. Working-class and poor parents regard it as their responsibility to shelter, feed, and clothe their children; teach them right from wrong; and comfort them. In these matters, Lareau reports, "language plays an important, practical role" (p. 139). But the parents did not focus on developing their children's language skills. They did not draw out their opinions or expect to be challenged by them. They disciplined their children with short, clear directives—sometimes coupled with physical punishment—which children generally accepted without complaint.

Institutions. Upper–middle class parents, Lareau finds, confidently engage institutions, and they teach their children to do the same. They are at ease with teachers, doctors, and others in authority, feeling free to ask questions and make demands, and they expect institutions to respond to their child's individual needs. In preparation for a wellness exam, an upper–middle class mother encourages her child to think of questions he wants to ask the doctor. "Don't be shy," she urges him, and he is not (p. 124). When another mother learned that her daughter had narrowly missed out on her school's gifted program, she sought advice from a network of well-informed friends, arranged to have her daughter retested, and prevailed on the school to assign her to the program. (Lareau's [2003] data show that high percentages of upper–middle class respondents, but few working class or poor respondents, know people who are doctors, lawyers, and psychologists [pp. 171, 285].)

Working-class and poor parents are, according to Lareau, intimidated by institutions and those who represent them. Mothers who have no difficulty making loud demands on the cable company are subdued in the presence of their child's teacher or doctor. They may not understand the words professionals use and do not feel competent to challenge their expertise. One mother whose fourth grader cannot read gets contradictory explanations from her child's teachers. But she makes no demands, leaving her daughter's education to the school's presumed experts. At the same time, passivity often masks an underlying resentment and distrust of middle-class institutions, which parents openly share with their children. They feel that such institutions operate according to an alien set of values. The parents of a fourth grader involved in playground conflicts encourage him to hit back, in defiance of the school's rules. The child is suspended. Working-class and poor parents who may use physical punishment fear having their children removed from the home if the school reports them to welfare authorities for child abuse.

Summarizing, Lareau suggests that the cultivated growth pattern encourages a sense of *entitlement* in upper–middle class children. Having been encouraged to participate in challenging organized activities, to speak freely with adults, to express their own opinions, to ask questions, to negotiate for what they want, and to expect institutions to respond to their needs, these children grow up with an enhanced sense of self-worth. They can be expected to deal confidently with institutions, which they see as sharing their own values.

In contrast, the natural growth pattern leads toward what Lareau characterizes as a sense of *constraint* in working-class and poor children. They have not been encouraged to cultivate formal language skills, to express their own opinions, or to question, challenge, or negotiate with adults. They have less experience than more privileged children with institutions, which they have been taught to regard with distrust.

Lareau sees her research as demonstrating the power of class to shape young lives. On the other hand, she concludes that race matters little, for children at this age, on the key dimensions of daily life, language, and relations with institutions. While black parents were inevitably concerned with the effects of racism on their young children, they differed little from white parents of the same class in their approach to child rearing.

To Lareau's surprise, the only class distinction that mattered was the one separating the child-rearing practices of the upper middle class from those of the two lower classes. Although she detected some differences between working-class and poor families, she found that they raised their children according to the same natural growth pattern. Perhaps a larger sample and the inclusion of families from the intermediate lower-middle class would have yielded a more complicated picture.

Bourdieu, whom Lareau cites as an inspiration for her work, would say that the upper–middle class children she studied were developing valuable cultural capital, which will serve them well as they move through the education system and into professional and managerial careers beyond. Kohn might add that the very character of upper–middle class occupations supports the cultivated growth pattern of child rearing, with its self-confident values. Both would agree that class differences in socialization support the reproduction of the class system.

School and Marriage

The class differences in patterns of socialization are reinforced by the tendency of children and adolescents to associate with others of like class background as they are growing up. Because neighborhoods tend to group people of similar economic means, the kids on the block are likely to be of the same social class. Local schools reproduce the class patterns of the neighborhoods they serve. Many upper-class and upper–middle class families make sure that their children's classmates will be from similar households by sending them to private schools. As Lareau's research shows, the upper–middle class pattern of scheduled activities outside of school reinforces class segregation.

Large public high schools in some communities bring together students of diverse backgrounds, but they do not necessarily mix freely. Often class differentiation within the school is institutionalized through curricula that separate students on the basis of their academic ability or postgraduation aspirations, which tend to be correlated with social class (Colclough and Beck 1986; Oakes 1985).

Students' own preferences contribute to the class patterning of association. A series of somewhat dated studies of adolescent cliques and friendships shows that students are inclined, but by no means certain, to choose friends who share their own class backgrounds (Cohen 1979; O. Duncan, Haller, and Portes 1968; Hollingshead 1949).

From 1988 to 2007, one of the authors of this book surveyed groups of students at a selective college regarding their associations in high school. Approximately 180 students were asked the occupations of the parents of their three best friends and most significant romantic interest in high school. Students also provided information on the occupation and income of their own parents. Families were stratified according to the Gilbert-Kahl model introduced in Chapter 1. The results are summarized in Table 5.2, which shows that the high school associations of this relatively privileged group of students were largely restricted to people with class backgrounds similar to their own but quite different from the class distribution of Americans generally. The first column shows the class distribution of all American households: Most (55%) are working class or below in our schema; relatively few (15%) are upper middle class or higher. As the second column indicates, the class distribution of students bound for an elite college is almost the reverse, with the large majority (71%) concentrated in the top classes. The third and fourth columns reveal that the students have formed friendships and romantic relationships with people who fit their own privileged class profiles. The strong pattern of class segregation among these affluent teenagers is, as we will see later in the chapter, consistent with the highly restricted association patterns of upper–middle class adults.

Like friendship, mate selection is influenced by social class. Sociologist Martin K. Whyte (1990) confirmed this conclusion from earlier studies in a survey of women in the Detroit metropolitan area. Whyte asked the respondents about the class positions of their parents and in-laws, at the time the women and their husbands were in high school. Given a choice of five classes, 58% of the respondents placed their parents and in-laws in the same class. Occupational comparisons between fathers and fathers-in-law produced similar results.

Table 5.2   Association Patterns of High School Students Bound for a Selective College

	Class Distribution (in percent)			
	U.S. National	*Students Surveyed*	*Students' Best Friends*	*Students' Best Dates*
Capitalist/Upper Middle	15	71	62	57
Middle	30	22	26	30
Working & Below	55	7	12	13
Total	100	100	100	100

SOURCE: Cumulative surveys of approximately 180 college students enrolled in social stratification course, 1989 to 2007.

In an earlier study of a Massachusetts community, Edward Laumann (1966) found that the inclination toward class endogamy (in-group marriage) was especially notable in the upper-middle class (Laumann's "top professional and business" grouping). Although this category constituted a relatively small proportion (24%) of the population surveyed, creating limited opportunities for endogamous marriages, 60% of the sons of top professionals and businessmen married women who shared their class background (pp. 74–81). A couple of quotations from Laumann's (1966) informants neatly reveal the attitudes underlying this phenomenon:

> What sort of husband would a carpenter be? What sort of education would he have? My viewpoint would not jibe with a carpenter. Marriage is based on equals. I would want my daughter to marry in her own class. She would go to college and would want her husband to be educated. I would want to be able to mix with in-laws and converse with them.
>
> I was born into a family of great privilege—but I think I have a responsibility. Family responsibilities will be better held if the background of the members are similar. My own interests are not manual. Our family relations are very close. It would be very demanding to have an unskilled or factory worker in the family. The tradition in our family is the professions or landowning. I would not turn out a daughter who married a machinist but—a machine operator is bound to have a modest education—probably not interested in intellectual things. Even with the best will in the world, a family relationship would be very difficult to accomplish. (P. 29)

The pattern of class endogamy observed by Laumann and others has an important implication for class differences in childhood socialization: If parents are from the same social class, they are likely to transmit a consistent set of class influences to their children.

Marriage Styles

Social class not only channels mate selection but also shapes the character of marital relationships. In general, sociological studies of husbands and wives have found clearer sex-role distinctions at lower class levels and greater intimacy, equality, and companionship at higher class levels (Langman 1987:222–234). A classic study conducted by Lee Rainwater (1965) supports this view.

Rainwater analyzed the marital role relationships of several hundred couples and distinguished three types of relationships, on a continuum: joint, intermediate, and segregated. *Joint relationships* focus on companionship and deemphasize the sexual division of labor. Husbands and wives with joint role relationships share the planning of family affairs, carry out many household duties interchangeably, and value common leisure activities. Even when responsibilities are parceled out by gender (wife–homemaker, husband–breadwinner), each partner is expected to take a sympathetic interest in the concerns of the other. In *segregated relationships,* there is clear differentiation of concerns and responsibilities, which minimizes the husband's involvement with household matters and the wife's with the world of (the husband's) work. Husband and wife are likely to have distinct leisure pursuits and separate sets of friends. Intermediate relationships fall between these two poles.

Based on answers to questions about family decision making, duties of husbands and wives, interests and activities of the partners, and the general character of the relationship, Rainwater placed each couple in one of the three categories. Of course, all couples were in some sense "intermediate." So the ratings were based on relative differences. When Rainwater compared the distributions of these three types of role relationships at four class levels, unmistakable differences emerged (Table 5.3). Joint relationships predominate in the upper–middle class (88%) and segregated relationships in the lower–lower class (72%), while the classes between them exhibit a neat gradient.

The relationship between social class and marital role types is more than a matter of academic curiosity. Reported marital happiness increases with class level, especially for women (Bradburn 1969: 156), and this phenomenon is tied to the character of the organization of marital roles. In Rainwater's study, middle-class couples reported greater sexual satisfaction than lower-class couples, but the difference was largely a function of the level of role segregation. For example, most lower-class wives in segregated relationships evaluated their sexual experience in marriage negatively, but lower-class wives in intermediate relationships were generally positive (Rainwater 1965: 28). In national surveys, companionship in marriage (which would appear to be similar to joint organization) is positively correlated with social class and with marital happiness (Bradburn 1969: 163).

Let's take a closer look at conjugal role types by examining how they function in upper–middle class and working-class families (L. Rubin 1976; Sussman and Steinmetz 1987: 226–231). In important ways, the very character of upper–middle class life lends itself to the joint role relationship. College life, generally a prologue to upper–middle class careers, delays marriage and encourages informal, relatively egalitarian association between men and women. High rates of social and geographic mobility are typical of this class. Husbands and wives are isolated from kin and removed from successive sets of friends as they move from community to

Table 5.3   Social Class and Conjugal Role Relationships

Class	Number	Role Relationships (percent)			
		Joint	Intermediate	Segregated	Total
Upper-middle class	(32)	88	12	–	100
Lower-middle class	(31)	42	58	–	100
Upper-lower class	(26)	19	58	23	100
Lower-lower class	(25)	4	24	72	100

SOURCE: *Family Design: Marital Sexuality, Family Size, and Contraception*, by Lee Rainwater. Chicago: Aldine. Reprinted by permission of Lee Rainwater.

community and up the career ladder. They must look to each other for support and companionship. Together, they are drawn into the career-oriented social life, such as entertaining clients or associates at home, that is one of the keys to success for ambitious executives and professionals.

Upper–middle class wives are expected to be "gracious, charming hostesses and social creatures, supporting their husbands' careers and motivating their achievements" (Kanter 1977:108). The traditional result has been the "two-person career" that links a husband's advancement to his wife's unpaid efforts. A more recent phenomenon, typical of younger couples, is the "dual career" family, in which both spouses pursue demanding professional or managerial careers. A survey of 1,000 working-age women in Chicago (Lopata et al. 1980) found that dual-career couples are as likely as single-career couples to mix social and professional life. About 60% of wives employed as managers or professionals reported that their husbands helped them with career-related entertaining at home. Husbands with professional or managerial jobs were somewhat more likely to receive such help from their wives. (It made little difference whether the wife was employed.) On the other hand, women employed in blue-collar jobs or married to blue-collar men reported little job-related entertaining.

As the Chicago study suggests, the career-oriented social life that becomes a shared endeavor for upper–middle class couples has no working-class equivalent. Working-class men and women do develop social ties on the job, but these tend to segregate rather than join husbands and wives. For example, many of the workers in a New Jersey chemical plant studied by David Halle (1984) drank together after work and joined coworkers on fishing trips and at sports events. Working-class occupations are less likely to require geographic mobility. Remarkably, most of Halle's chemical workers were born within 2 miles of the plant where they worked (p. 303). Under such circumstances, it is easier for spouses to maintain ties with kin and friends from adolescence and early adult years. Dependency on the couple's parents is intensified by the economic insecurity that is especially typical of young working-class families. These social ties tend to draw husband and wife to separate sources of support and companionship outside the marriage.

Bott's (1964) work in England showed that couples who come to a marriage with tight-knit networks of friends and kin (that is, the people each spouse knows tend to know one another) and maintain these ties are the most likely to develop segregated marital relationships. Her data suggest that social networks of that sort are least typical of professionals and most typical of manual workers.

We have dealt with the origins of joint and segregated role relationships in experiences typical of the top and bottom of the class order. What can we say about the mix of marriage types Rainwater found in the middle of the class structure (Table 5.3)? Two social factors seem relevant. One is social mobility: People moving up or down in the class structure may carry with them lifestyles acquired in their class of origin. Thus, the upper–middle class origins of many lower–middle class couples (especially younger couples) can help explain the predominance of joint relationships among them. An analogous argument can be made for the spread of segregated relationships upward. The second factor is cultural: the tendency of upper–middle class lifestyles to become generally fashionable models and filter downward.

Through these processes, couples are exposed to conflicting influences, which may be reflected in intermediate role relationships.

Sex-role socialization is another source of class differences in marital role organization. Kohn's research (1969), which we touched on earlier in this chapter, found that working-class parents are more likely than middle-class parents to hold separate sets of expectations for boys and girls. A study of college-age women (Vanfossen 1977) found that college-age daughters of working-class fathers are more likely than their middle-class peers to subscribe to traditional sex-role values as expressed in questionnaire items such as "A woman should not expect to go to exactly the same places or have the same freedom as a man." Such women are the most likely to find the segregated marital role acceptable. Boys of all classes are taught to be more controlled, more instrumental, less emotional, and less empathetic than their sisters, but the distinction is made much more emphatically in blue-collar families. Lillian Rubin (1976), a sociologist and psychotherapist who conducted lengthy interviews with working-class and upper–middle class couples, noted big differences in the behavior of their sons:

> Not once in a professional middle-class home did I see a young boy shake his father's hand in a well-taught "manly" gesture as he bid him good night. Not once did I hear a middle-class parent scornfully—or even sympathetically—call a crying boy a sissy or in any way reprimand him for his tears. Yet, these were not uncommon observations in the working-class homes I visited. Indeed, I was impressed with the fact that, even as young as six or seven, the working-class boys seemed more emotionally controlled—more like miniature men—than those in the middle-class families. (P. 126)

Boys who are taught to be "manly" in this way are less likely as adults to feel comfortable with joint role relationships in marriage.

Blue-Collar Marriages and Middle-Class Models

Two intimate studies of working-class life, Rubin's book and another by E. E. LeMasters (1975), published about the same time, cast further light on the differences between working- and middle-class marriages. Rubin, who interviewed young parents in their Northern California homes, and LeMasters, who spent 5 years getting to know the somewhat older patrons of the Oasis, a "family-type" working-class tavern in Wisconsin, reached surprisingly similar conclusions. Both found that marital norms filtering down from the upper-middle class were creating enormous strains in blue-collar marriages.

One of LeMasters' (1975) informants, a woman married for 30 years, bitterly described the traditional pattern of segregated blue-collar marriages:

> The men go to work while the wife stays home with the kids—it's a long day with no other adult to talk to. That's what drives mothers to the soap operas—stupid as they are.

Then the husband stops at some tavern to have a few with his buddies from the job—not having seen them since they left to drive home ten minutes ago. The poor guy is lonely and thirsty and needs to relax before the rigors of another evening before the television set. Meanwhile the little woman has supper ready and is trying to hold the kids off "until Daddy gets home so we can all eat together." After a while, she gives up this little dream and eats with the kids while the food is still eatable. About seven o'clock, Daddy rolls in, feeling no pain, eats a few bites of the overcooked food, sits down in front of the TV set, and falls asleep.

This little drama is repeated several thousand times until they have their twenty-fifth wedding anniversary and then everybody tells them how happy they have been. And you know what? By now they are both so damn punch drunk neither one of them knows whether their marriage has been a success or not. (P. 42)

Such dissatisfaction was probably nothing new, but as the tone of her comment suggests, expectations were changing. By the 1970s, the traditional pattern was being challenged by notions of intimacy, companionship, sharing, and equality received from above. The problem was and still is that these ideals do not appeal equally to wives and husbands. Women are prepared for them by their socialization and in many cases by contact with a middle-class world through white-collar employment. They are exposed to the newer conceptions of marriage through women's magazines, television soap operas, and daytime talk shows. Men, on the other hand, are likely to be satisfied with established role relationships, which they have regarded as part of the natural order of things. The traditional women's role is, according to one of LeMasters' informants, "natural for them so they don't mind it" (p. 105). Men sense, of course, that many women do "mind it," but they are inclined to think that women's complaints are groundless. At the Oasis, a construction worker asks LeMasters,

What the hell are they complaining about? My wife has an automatic washer in the kitchen, a dryer, a dishwasher, a garbage disposal, a car of her own—hell, I even bought her a portable TV so she can watch the goddamn soap operas right in the kitchen. What more can she want? (P. 85)

But behind the bluff, there is fear. From the less "macho" setting of his living room, one of Rubin's (1976) informants phrased the problem differently:

I swear I don't know what she wants. She keeps saying that we have to talk, and when we do, it always turns out I'm saying the wrong thing. I get scared sometimes. I always thought I had to think things to myself; you know, not tell her about it. Now she says that's not good. But it's hard. You know, I think it comes down to that I like things the way they are, and I'm afraid I'll say or do something that'll really shake things up. So I get worried about it, and I don't say anything. (P. 121)

For their part, working-class women in these studies are very dissatisfied but also frightened and confused and occasionally given to wondering whether asking a man to be more than a conscientious provider is indeed asking too much.

> I'm not sure what I want. I keep talking to him about communication, and he says, "Okay, so we're talking, now what do you want?" And I don't know what to say then, but I know it's not what I mean. I sometimes get worried because I think maybe I want too much. He's a good husband; he works hard; he takes care of me and the kids. He could go out and find another woman who would be very happy to have a man like that and who wouldn't be all the time complaining at him because he doesn't feel things and get close. (Rubin 1976:120)

A second aspect of blue-collar marriage was under strain in the 1970s: sexual adjustment. Problems in this area were not new. For instance, husbands and wives had long clashed over the desirable frequency of sexual intercourse. LeMasters (1975) heard this complaint among patrons of the Oasis (p. 101). However, difficulties of more recent origin, deriving from the sexual revolution of the 1960s and 1970s, were evident in the comments of the younger couples interviewed by Rubin. In the 1970s, working-class sexual behavior was moving closer to middle-class norms. For example, working-class couples had nearly caught up with middle-class couples in their willingness to engage in once-exotic sexual variants such as cunnilingus and fellatio; working-class men had become similar to middle-class men in their concern for their wives' sexual satisfaction (Rubin 1976:134–135, 137–148). But change has psychological costs. Again, differential receptivity to new standards was creating stress for working-class marriages. In this case, men were more open to change. The blue-collar workers Rubin interviewed wanted freer, more expressive, more mutually satisfying sexual relationships with their wives, as their remarks show:

> I think sex should be that you enjoy each other's bodies. Judy doesn't care for touching and feeling each other, though. She thinks there's just one right position and one right way—in the dark with her eyes closed tight. Anything that varies from that makes her upset. It's just not enjoyable if she doesn't have a climax, too. She says she doesn't mind, but I do. (P. 136)

Rubin and LeMasters portrayed the powerful influence of upper–middle class models on working-class behavior, even in intimate realms of experience like marital gender roles and sexuality. But has the behavior of working-class men and women continued to change along the path they described in the 1970s? Two more recent books contain some clues: Halle's (1984) study of chemical plant workers referred to earlier and Lillian Rubin's recent book on working-class life in an age of economic insecurity (1994).

Halle observes that most of the men he studied endorse the modern marital ideals of companionship and mutually fulfilling sexuality. But, by and large, these ideals were not reflected in their own conflict-ridden marriages. Halle stresses that husbands and wives typically do not share leisure interests, and the men are drawn to a male leisure culture that excludes women and absorbs much of men's time. The following complaints, the first from a wife and the second from one of the male chemical workers, are typical:

> I never see Jimmy. He's out a lot, and when he's home he likes to watch sports on TV. And I hate sports, especially football.

> Susan keeps saying, "Why don't we go away together? Just you and I." But that's boring. I like to go fishing and drink a few beers when I'm on vacation, but . . . she doesn't like to fish, and she thinks I drink too much. (Halle 1984:55–56)

Such comments suggest differences more deep-seated than incongruent leisure interests. In fact, only one-third of Halle's respondents claim to be happy with their marriages. This minority does stress the importance of companionship. One worker tells the author,

> I don't care what they [the other workers] say, marriage is a beautiful thing. But you have to work at it. You can't just leave your wife alone like a lot of these guys do. . . . I'll tell you what makes a good marriage. It's when you do things together. (Halle 1984:56)

Rubin's book (1994), published two decades after her original study, portrays a very different world. The lives of working-class women have been transformed, reflecting changed conceptions of womanhood in the broader culture and the new economic role of women. In the 1990s, as we noted in the last chapter, wives and mothers are more likely to hold jobs and to work full time. As the real wages of working-class men have declined, their families have become increasingly dependent on the earnings of women. People have abandoned the notion—widely held in the 1970s—that men could and should support families by themselves or that women's earnings were merely supplementary to the household budget.

Rubin (1994) reports that the expanded economic role of women is altering expectations about the domestic roles of husbands and wives. Working-class women are now more likely to demand equality and reciprocity from their husbands. Even the relatively traditional young secretary who rejects "feminism" and says a woman should make her man "feel like a king," tells Rubin that she will not "bow down" to a man or be like her mother who "waits on my father all the time." "Why should I?" she asks. "I'm as equal as him" (p. 74). The attitudes of men are also changing. In the 1970s, Rubin (1994) interviewed few working-class husbands "who would even give lip service to the notion of gender equality" (p. 77). By the 1990s, such views were common, especially among younger men. But Rubin, like Halle, found a wide gap between professed ideals and everyday behavior. In particular, she notes that the household division of labor has been slow to change and has become the subject of growing contention between husbands and their hard-pressed working wives.

Rubin also interviewed the teenage children of the working-class families she studied in the 1990s. Among the daughters especially, a transformation is evident. Their mothers dreamed of little but marriage. They married young because they were pregnant or anxious to escape the smothering authority of traditional families.

Now the daughters are more independent and have wider horizons. They also have fewer illusions about what young men with limited education can now provide. They expect to marry "someday," but first they want to work, to live on their own, to travel and experience the world. One daughter, single and pregnant like her mother before her, gets an abortion rather than follow her mother's path into a bad marriage. When they do marry, these young women will likely be very different from their mothers.

The studies we have examined in this section suggest that working-class marriage has been changing over the last generation, under economic and cultural pressure. At least at the level of values, working-class couples are moving toward upper–middle class models. But without new studies that make systematic class comparisons, like Rainwater's survey (1965) or Rubin's original book (1976), we cannot be sure of the current extent of class differences. Both classes are affected by the changing roles of women and the transformation of the economy, but they may be reacting to these changes in different ways.

Social Class and Domestic Violence

In *The Unknown City: The Lives of Poor and Working-Class Young Adults*, sociologists Michelle Fine and Lois Weis (1998) deal at length with a topic that is often slighted in the literature on class and family life: domestic violence. Fine and Weiss conducted in-depth interviews in the early 1990s with an ethnically diverse group of young men and women (age 25 to 35) in Buffalo, New York, and Jersey City, New Jersey. Most of their respondents would seem to fit into our "working poor" and "underclass" categories (see Figure 1.1, p. 13). Living in two deindustrialized cities in a period of high unemployment and low wages for less-educated workers, they are very much the victims of the Age of Growing Inequality.

Domestic violence is a persistent theme in the researchers' interviews with the young women—white, black, and Hispanic—in their sample. One respondent recalls her childhood as follows:

> [T]here was blood in our house just about every day. Somebody was always wacked with something. And dinner, to this day, I don't sit and eat dinner with my kids. We eat in the parlor in front of the TV, or whatever. Because every time we sat and ate . . . a fight broke out and, you couldn't leave the kitchen. So you had to sit there and listen to it. (Fine and Weiss 1998:143)

Another young woman describes a brutal, chaotic relationship with a boyfriend who would beat her regularly and then claim,

> it was my fault. I made him do it because I yelled at him and he couldn't handle it. . . . I'd block the door and he'd kick the door right in. . . . And this went on for a year. I told him the next time you hit me, don't sleep here, 'cause I will chop you up. I sat in a chair with an ax in my hand and said I was gonna chop him up that night. (P. 151)

In such lives, "not getting beat up" is one mark of a good relationship. A 21-year-old mother offers this assessment of her current relationship with her fiancé, the father of her son:

> It's good; I mean, he's there for me. It's good. I don't know what to say (laughs). . . . He listens to me. He's a friend. I don't know, I guess I got all the conveniences of a nice relationship. . . . I don't get beat up; I don't get put down. (Pp. 153–154)

Fine and Weis (1998) observed that domestic violence can be found in all social classes, but they emphasized the high rates and intergenerational character of family violence at lower class levels. The lives of poor and working-class women, they write, "are saturated with domestic terror" (p. 134). It is possible that these young adults in Buffalo and Jersey City, drawn from the poor and, it appears, the lower fringe of the working class, during a period of high unemployment, represent an extreme. Young adults are the most prone to violence in relationships, and especially so in difficult economic times.

Rubin (1994), whose work we reviewed in the last section, did the research for her second book during the same period, with a sample that was, on average, probably a little better off than the Buffalo and Jersey City respondents. Fourteen percent of her families acknowledged domestic violence, but the true figure could be higher, she writes, since, "this is one of the most closely guarded secrets in family life" (p. 116). A teenage boy Rubin interviewed refused to join the conspiracy of silence: "I bet they didn't tell you he beats my mother up, did they? Nobody is allowed to talk about it; we're supposed to pretend like it doesn't exist" (p. 116).

Rubin (1994) found that men were especially likely to become abusive in periods of unemployment. One respondent, who said he had not abused his wife but feared he might, shared his feelings with Rubin:

> It's hard enough being out of work, but then my wife gets on my case, yakking all the time about how we're going to be out on the street if I don't get off my butt, like it's my fault or something that there's no work out there. When she starts out like that I swear I want to hit her, anything to shut her mouth. (Pp. 115–116)

We can gain a more systematic picture of class patterns in domestic violence by looking at statistics on violence against women by "intimate partners" drawn from the U.S. Justice Department's (2001) National Crime Victimization Survey. "Intimate partners" as defined by the survey includes current and former husbands and boyfriends.[2] The great advantage of this survey is that it collects data anonymously and includes both crimes reported and not reported to the police.

---

[2]Violent "intimate partners" may also be females, but it is unlikely that there are enough offenders in this category to affect the statistics. The crimes covered include simple and aggravated assault, sexual assault, rape, robbery, and murder.

In 1999, according to the Justice Department (2001), females 12 and older were victims of intimate partner violence at a rate of 6 per 1000. But this low figure (less than 1%) conceals enormous demographic variation. It is also an annual rate; individual experiences cumulated over decades or a lifetime would produce a much higher incidence of victimization. The Justice Department also reports that the rate dropped 41% from 1993 to 1999, a period during which unemployment decreased sharply—a trend consistent with Rubin's emphasis on the connection between economic stress and domestic violence.

Figure 5.1 reports annual rates of intimate partner violence by age and income level. Two facts stand out in sharp relief. First, victimization is most frequent between the ages of 16 and 49, especially between 16 and 34. Second, there are enormous differences by income level. Women between 16 and 35 at the lowest income level are three to five times more likely to be victims of intimate partner violence than their peers in the top income category. These data support the claims of Fine and Weis, among others, that domestic abuse is much more common at lower class levels.

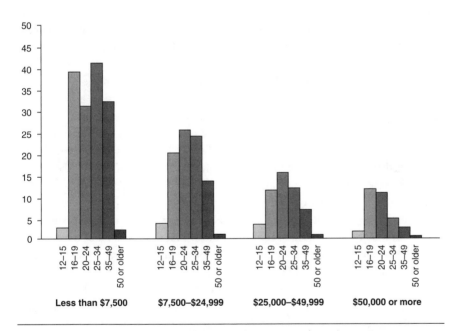

Figure 5.1   Social Class and Violence Against Women by Intimate Partners

SOURCE: U.S. Department of Justice 2001.

Informal Association Among Adults

Warner, whose classic Yankee City study we examined in Chapter 2, considered patterns of association so critical to understanding the class system that he sometimes appeared to define class in terms of association. A social class, he suggested, is a group of people who belong to the same social cliques, intermarry, dine in each other's homes, and belong to the same organizations.

Warner defined a social clique as "an intimate nonkin group," with no more than 30 members. Warner's research team collected elaborate data on the clique membership of families in Yankee City. They found that most cliques joined people of the same or adjacent classes (Warner and Lunt 1941:110–111, 350–355).

The notion that social class is about "who you hang out with" is widely shared. Asked to discuss the basis of social class differences, almost half of the skilled blue-collar workers and two-thirds of the white-collar workers in a Providence, Rhode Island, study referred to patterns of association. Their comments suggest that people belong to the same class if they "run around together"; intermarry; "belong to the same churches, clubs, organizations"; "live in the same neighborhoods"; or send their kids to the same schools (Mackenzie 1973:148).

Numerous studies suggest that patterns of association are shaped by social class—though association is also influenced by factors that cut across class lines, including gender, race, age, religion, and shared interests (Allan 1989; Argyle 1994:6692; Smith and Macaulay 1980). Earlier in this chapter, we saw that adolescent friendships and marital mate selection reflect class backgrounds. Adult friendships are also patterned by class, according to surveys done in Providence, the Boston area, and metropolitan Detroit (Laumann 1966, 1973; Mackenzie 1973).

The findings of Laumann's (1966) survey of 422 white men in the Boston area are typical. Laumann asked 422 white male residents of Cambridge and Belmont the occupations of their three closest friends. He stratified his sample into five occupational classes. Laumann found, for example, that three-quarters of the friends of men in his top class (professionals/higher business) were from the same class and three-quarters of the friends of lower-skilled blue-collar workers were from the same class or the adjoining-skilled blue-collar class. He also concluded that the blue-collar/white-collar line was a barrier to friendship. Only 25% of all close friendships in the study were between blue-collar and white-collar workers.

Social class, then, channels friendship choices. Research shows that it also influences the extent and character of informal association. The literature suggests that people at higher class levels (1) have more friends and more active social lives; (2) are less likely to preserve friendships from their youth; (3) spend proportionately less time with relatives; (4) are more likely to entertain friends at home and, in particular, to host dinner parties; (5) are more inclined toward couple-oriented social activities; (6) are more likely to develop (nonromantic) cross-sex friendships; and (7) are more likely to mix career and social life.[3]

How can we explain the class patterning of informal association? Why do people tend to marry and maintain friendships with class peers? What accounts for the class differences in the character of social life? Two obvious but powerful factors are money and propinquity (physical or social proximity).

---

[3]Allan 1989; Argyle 1994:66–92; Curtis and Jackson 1977:169; Dotson 1950; Kahl 1957:138; Kanter 1977; Rubin 1976, 1994; Shostak and Gomberg 1964; W. Whyte 1952.

Dinner parties can be costly affairs, and guests are expected to reciprocate in kind. Skiing and sailing are more expensive than bowling. These price-of-admission differences segregate leisure activities and the people who engage in them by ability to pay. In everyday life, people tend to encounter others who are close to them in status. They live in neighborhoods and send their children to neighborhood schools that are relatively homogeneous in household income. Their coworkers have similar jobs—except for bosses, whose authority places them at a social distance. In short, daily life is structured in ways that limit the opportunities to develop social ties across class boundaries.

Beyond money and propinquity are a series of more subtle factors—matters of prestige, style, interests, values, and comfort level. Their influence on patterns of informal association is suggested by some of the comments of Laumann's (1966) respondents. One man indicates that he has "nothing in common" with people in lower occupations, another characterizes factory workers as "rough," and a third complains about the "uppity" attitudes of a relative who is a successful executive (pp. 28–29). These men seem uncomfortable with disparities in social prestige. But their attitudes also reflect objective differences in areas including education (which is strongly correlated with class) and occupational experience. Education produces contrasts in language usage, attitudes, and personal interests. Adults with limited education are likely to be uneasy in the presence of the well educated.

Occupational experience, as Kohn's studies of socialization demonstrate, contributes to class differences in values. Halle (1984) emphasizes that working-class people typically have dull "jobs," while upper–middle class people have engaging "careers." The former are inevitably less interested in conversations that revolve around work and generally less inclined to mix work and leisure.

Formal Associations

Like informal ties, participation in formal associations is patterned by social class.[4] Formal associations are large groups or organizations with explicit purposes and rules of membership, including the YMCA, the Elks Club, the Teamsters union, the Burning Tree Country Club, and the Boy Scouts. From its beginnings, the United States has been characterized by observers as a nation of joiners. Today, this generalization is somewhat less than half true. Most working- and lower-class Americans have little or no participation in formal associations. Even the participation of the lower–middle class is modest. The true joiners are members of the upper-middle and upper classes, who are especially likely to participate in civic and charity organizations.

Members of these top classes are not just the most likely joiners. They are also the most active participants in organizations and, even when organizational membership cuts across classes, the most likely to serve in leadership positions. The

---

[4]Hodges 1964:105–115; Mackenzie 1973:81–84; Smith and Macaulay 1980; Warner et al. 1949a.

reasons for this phenomenon are not hard to imagine. These managers and professionals enjoy the prestige attached to high-class position. They have more education. At work, they develop organizational skills and confidence as leaders. Finally, many see active participation in community organizations as a way to bolster their careers.

Associations often draw their membership from a limited range in the class structure. Country clubs and exclusive social clubs such as New York's Links or Boston's Sommerset draw from the upper and upper–middle classes. Service organizations such as Rotary or Lions, fraternal orders like the Elks Club, and patriotic organizations like the Veterans of Foreign Wars recruit members from successively lower class levels.

Even churches—institutions supposedly rejoicing in our common humanity—are class typed. According to the existing research, people of higher status are likely to attend those Protestant denominations that feature services of quiet dignity and restrained emotion, such as the Episcopal or Unitarian groups. Middle-status people are more often seen at the Methodist, Mormon, and Lutheran churches. Lower-status individuals are most likely to join revivalist and fundamentalist churches, such as the Pentecostals. The class level of Catholic congregations seems to vary with the ethnicity of the congregation, reflecting the timing of their immigration to the United States.[5]

Separate Lives

Americans are increasingly segregated by social class. No one has made this point better than Tom Wolfe in his novel *Bonfire of the Vanities* (1987). The novel's protagonist, Sherman McCoy, is a smug, young Wall Street trader with a $3 million Park Avenue apartment and few redeeming qualities.

Driving into the city in his Mercedes one evening, accompanied by his mistress, Sherman blunders into an impoverished ghetto neighborhood, where he is involved in a fatal hit-and-run accident. Subsequently, McCoy finds himself locked up in a courthouse holding cell in the unwanted company of dozens of tough young men—poor and dark-skinned like the victim of his Mercedes. These events initiate a downward spiral in McCoy's life. By the end of the novel, a year after the accident, McCoy is separated from his wife, his mistress, his money, and his lawyer, who has resigned from the case because McCoy is broke.

The action of *Bonfire of the Vanities* is driven by its satisfying but improbable premise: Sherman McCoy has smashed through the wall that normally separates privileged people like the McCoys from people like the accident victim and Sherman's cellmates—or for that matter, from the $36,000-a-year assistant D.A. who prosecutes the case. "If you want to live in New York," a friend advises McCoy, "you've got to insulate, insulate, insulate" (Wolfe 1987:55). Before the accident, McCoy used his money to do just that. His world was as insulated from the grimy reality of the city as the posh cabin of his Mercedes from the asphalt below.

---

[5]Demerath 1965; Laumann 1966:55; Smith and Macaulay 1980:514; Kosmin and Lachman 1993:257–269).

Wolfe's novel reflects the growing disparities of the current era. It portrays a society whose members, divided by class and race, live in increasing isolation from one another; they no longer share (despite McCoy's strange fate) a common destiny. Journalist Mickey Kaus develops this theme in *The End of Equality* (1992), arguing that rising economic inequality has been accompanied by rising social inequality. Money, fear of the poor, and an inflated sense of their own superiority are motivating prosperous Americans to develop separate lives. "An especially precious type of equality—equality not of money but in the way we treat each other and live our lives—seems to be disappearing" (p. 5).

Kaus looks back at the post–World War II era as (with the "evil" exception of race) "a golden age of social equality." The war, perhaps more than any event in our history, provided Americans with a common experience and a sense of shared destiny. Wealthy 26-year-old John F. Kennedy served on a small PT boat in the South Pacific with men who had been machinists, factory workers, truck drivers, and night school students. Seventy percent of able-bodied young men, most of them drafted, served in the military (Kaus 1992:50). Some, like Kennedy's brother Joe, did not survive. Those who returned brought with them a network of friendships, forged under the threat of death, with little regard for class differences.

After the war, the GI Bill offered veterans of diverse backgrounds access to college and the opportunity to buy their own homes. In the midst of the shared prosperity of the 1950s, there was a sense that the social distance between Americans of different classes was shrinking. Today, it appears to Kaus that just the opposite is happening. Against a backdrop of growing economic inequality, Americans worry about the emergence of what they take to be a permanent underclass. The opulent lives and social pretenses of the rich—objects of ridicule in a more egalitarian age— inspire fawning articles in glossy magazines aimed at upper–middle class readers in search of role models. Professionals with merely comfortable incomes see themselves as "not just richer, but more civilized, better educated, wittier, smarter, cleaner, prettier" than the average American (Kaus 1992:27).

"Who killed social equality?" asks Kaus. Oddly, he rejects the most obvious suspect, rising economic inequality, and insists that the guilty party is "the decline of the public sphere." What he has in mind is the reduction of the social realm where Americans of different classes meet on more or less equal terms. He points to the end of the draft and the replacement of a broad-based citizen military with a volunteer force, which recruits few soldiers from the upper end of the class structure. But most of his examples revolve around residential segregation by class. As the rich and relatively rich retreat to exclusive suburbs (sometimes even to "gated private communities"), they separate themselves from the less privileged. Here, public spaces—the mall, the supermarket, the drugstore, the coffee shop—are largely inhabited by other members of the privileged classes. There is no place like the bar portrayed in the TV sitcom *Cheers*, a democratic setting where the postman and the psychiatrist meet informally. Above all, children attend school with others of the same class, even if they do not enroll in private academies. And, not surprisingly, their parents, who support budget-slashing politicians, do not hesitate to vote for local school taxes. They know their own kids will benefit.

Journalist Kaus is examining the same fractured social reality as novelist Wolfe. Has he proved anything? Is the social distance between Americans of different classes actually increasing, and is it doing so for the reasons that Kaus suggests? *The End of Equality* is, unfortunately, full of vivid examples but short on systematic evidence. This chapter has examined a series of studies demonstrating that Americans are, in fact, separated by class in their everyday lives. But the attentive reader will have noted that many of the studies are rather dated. There are few recent studies of class patterns of association and hardly any that would allow dependable comparisons with the past. (Of course, we need such comparisons to think seriously about the issues of change Kaus raises.) There is, however, an important category of exceptions: the studies of changing residential patterns based on the decennial U.S. Census. By comparing the geographic distribution of households at various class levels from one census to the next, we can determine whether class segregation is actually increasing.

Figure 5.2, based on analyses of census data by Massey and Fischer (2003), traces changes in the class segregation of neighborhoods (census tracts) in metropolitan areas from 1970 to 2000. The generally upward trajectories of the three curves demonstrate that class segregation has increased in the Age of Growing Inequality. The chart measures the degree to which families with high and low incomes live in the same neighborhoods, using an index of dissimilarity that runs from 0 to 1.0. An index of 1.0 would indicate that the affluent and the poor are absolutely

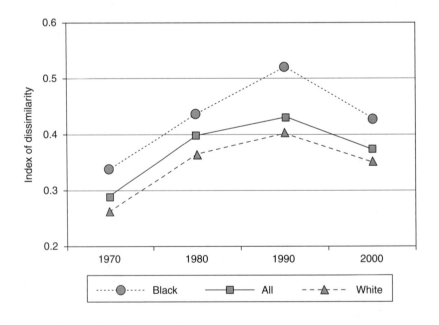

Figure 5.2   Residential Segregation by Income, 1970–2000

SOURCE: Based on Massey and Fischer 2003:27.

segregated—none live in the same area; an index of 0 would indicate that they are perfectly integrated.[6]

The chart shows that residential segregation by income rose about 50% from 1970 to 1990, the period when the effects of deindustrialization were felt in many metropolitan areas, and then dropped off slightly in the 1990s, a decade, especially after 1995, of relative prosperity. The chart also reveals that blacks are much more income segregated than whites. These findings provide at least some support for Kaus's argument about growing social inequality, which emphasizes residential separation. It appears that even Americans of relatively modest affluence are heeding the advice of Sherman McCoy's friend: "Insulate, insulate. . . . "

Conclusion

This chapter has explored the social implications of class structure, emphasizing socialization and association. The life cycle has served as a guiding thread. We learned that children are precociously aware of class distinctions and that they are socialized according to patterns that reflect the class position of their parents. Adolescents tend to form friendships and romantic ties with others who share their class background. Young adults typically marry class equals or near equals. Adult friendships, marital styles, residential choices, organizational activities, and even church membership are all patterned by social class.

These observations bring us back to Max Weber's idea, invoked at the beginning of the chapter, that prestige classes (or "status groups," as he called them) are social "communities," characterized by distinctive lifestyles and values. To a remarkable extent, our social lives and outlooks are molded by class position. Consider the typical differences we have found between the upper–middle and working classes. Members of the upper–middle class generally share the life-shaping experiences of college, late marriage, geographic mobility, and a career-oriented social life. They are drawn to joint models of marriage and values of self-direction and tolerance, which they stress for their children. Their child-rearing practices inculcate formal language skills, confidence in dealing with institutions, and a general sense of entitlement. Their exclusive choices of friends and mates suggest that this class is relatively isolated from the rest of the population. In contrast, members of the working class tend to get married younger, retain stronger ties with kin and friends from adolescence and young adulthood, and maintain stronger separation between social and economic life. They are likely to develop role-segregated marriages and to inculcate values of discipline and conformity in their children. Their child-rearing practices convey a sense of constraint to their sons and daughters.

---

[6]Low ("poverty") income was defined by the federal poverty line and varies with family size. High ("affluent") was defined as four times the poverty level. The income thresholds were recalculated for each census year to account for inflation. In 2000, the figures were, respectively, $17,029 and $68,116 for families of four.

These differences reinforce Bourdieu's conclusion that the advantages of the privileged classes extend beyond economic capital to cultural and social capital. The upper–middle class child who grows up with superior command of the English language and has learned to deal confidently with people in authority has accumulated valuable cultural capital. Upper–middle class parents, whose friends and relatives typically include doctors, psychologists, lawyers, and other professionals, possess valuable social capital they can tap when they need advice.

Are these social differences between classes becoming more or less pronounced? We might expect social differences to widen as economic differences grow, and we know that residential segregation by class has intensified since 1970. But we have little other evidence about recent trends. Whatever the direction of change, the American class system is far from becoming a series of discrete class cultures. The differences we have described in this chapter are statistical tendencies, not absolute contrasts. Americans of all classes are influenced by a national culture and share many key values. They are exposed to the same ideas and lifestyles in pervasive mass media. At the same time, the diversity of American society and (as we see in the next chapter) relatively high rates of social mobility guarantee that the membership of any social class will be quite varied.

Key Terms Defined in the Glossary

association	lifestyle
blue-collar workers	segregated marital relationships
cultural capital (see capital)	social capital (see capital)
economic capital (see capital)	social clique
endogamy	socialization
joint marital relationships	white-collar workers

Suggested Readings

Argyle, Michael. 1994. *The Psychology of Social Class.* New York: Routledge.
  Summary of social-psychological literature on class. Areas covered include social relationships, lifestyles, and personality differences. Extensive bibliography.

Bourdieu, Pierre. 1984. *Distinction: A Social Critique of the Judgement of Taste.* Cambridge, MA: Harvard University Press.
  Class cultures and their function in defining and reproducing class differences.

Bourdieu, Pierre. 1986. "The Forms of Capital." In *Handbook of Theory and Research for the Sociology of Education,* edited by John Richardson. New York: Greenwood Press.
  A compact discussion of the economic, cultural, and social forms of capital.

Brooks, David. 2000. *Bobos in Paradise: The New Upper Class and How They Got There.* New York: Simon & Schuster.
  The origins, lifestyles, and values of the emerging, knowledge-based privileged class. Witty and insightful.

Hollingshead, August B. 1949. *Elmtown's Youth.* New York: Wiley.

> *This classic work from the Warner era includes a comprehensive study of patterns of association among high school students. Demonstrates strong correlation between social class and social cliques.*

Kingston, Paul W. 2000. *The Classless Society.* Palo Alto, CA: Stanford University Press.

> *Challenges the notion of discrete classes and the value and lifestyle distinctions associated with them. Critical examination of empirical evidence.*

Kohn, Melvin and Carmi Schooler. 1983. *Work and Personality: An Inquiry into the Impact of Social Stratification.* Norwood, NJ: Ablex.

> *Synthesis of their research.*

Lamont, Michele. 1992. *Money, Morals, and Manners: The Culture of the French and the American Upper-Middle Class.* Chicago: University of Chicago Press.

———. 2000. *The Dignity of Working Men: Morality and the Boundaries of Race, Class, and Immigration.* Boston: Harvard University Press.

> *The moral boundaries that working-class and upper–middle class men draw between themselves and others. Both based on in-depth interviews conducted in the United States and France.*

Langman, Lauren. 1987. "Social Stratification." In *Handbook of Marriage and the Family,* edited by Marvin Sussman and Suzanne Steinmetz. New York: Plenum.

> *Wide-ranging essay summarizing literature on class differences in family life.*

Lareau, Annette. 2003. *Unequal Childhoods: Class, Race, and Family Life.* Berkeley: University of California Press.

> *An original and influential study of class differences in child rearing.*

Rubin, Lillian Breslow. 1976. *Worlds of Pain: Life in the Working-Class Family.* New York: Basic Books.

———. 1994. *Families on the Faultline: America's Working Class Speaks About the Family, the Economy, Race, and Ethnicity.* New York: HarperCollins.

> *Sensitive portrayals of working-class life. First volume makes revealing comparisons with upper-middle class.*

CHAPTER 6

Social Mobility

The Societal Context

*Some people's money is merited
And other people's is inherited.*

Ogden Nash, *The Terrible People*

A rising tide lifts all boats.

John F. Kennedy

S ocial mobility, our topic in this chapter and the next, may be defined as the extent to which people move up or down in the class system, especially from one generation to the next. The study of social mobility is motivated by curiosity about a seemingly simple question: is our place in the class system, in Ogden Nash's words, merited or inherited?

We will look at mobility from two perspectives, societal and individual. The first, the focus of this chapter, considers the system as a whole, emphasizing the general pattern of social mobility. We ask this: How much mobility is there in America? Are the opportunities for mobility increasing or decreasing? What factors are shaping mobility? In Chapter 7, we look at individuals. We ask why, given the available opportunities, some move ahead, some fall behind, and others just stay in place. In particular, we want to know how much importance to attach to family background factors, like parental occupation and race, and how much to an individual's educational attainment.

The keys to the societal perspective are the shape of the class structure and its relative openness. Consider, for example, a hypothetical feudal society made up of a mass of peasants dominated by a handful of large landowners and their overseers. Its structure is that of a very steep pyramid: broad at the base, narrow in the middle, and needle-pointed at the top. Even if the society were perfectly open or "fair" and gave everyone an equal opportunity in the competition for the higher positions,[1] the chances of getting ahead would be slim. The children of peasants would almost certainly grow up to be peasants. If opportunities were truly equal, most of the children of aristocrats would also become peasants.

Now suppose that this society begins to industrialize. A new class of urban capitalists emerges at the top. Its activities promote the growth of an industrial working class. The middle sectors expand because an industrial society requires rising numbers of engineers, accountants, managers, teachers, electricians, and other specialists to function efficiently. The social pyramid is becoming fatter in the middle and somewhat wider at the top. Even if this society does not treat everyone equally, the chances of moving up are now much greater. The children of peasants might become factory workers (an improvement over farm work), and their grandchildren could aspire to even higher positions. In short, there is a quickening of upward social mobility due to structural change.

How Much Mobility?

A common way to measure intergenerational mobility is to compare the occupations of fathers and sons, as we have done in Tables 6.1 and 6.2. (We will look at the mobility of women in the next section.) These tables are based on recent national surveys that ask respondents their occupations and the occupations of their fathers

---

[1]Of course, historically, feudal societies were far from open or fair.

when they were growing up.[2] Using this information, we can answer questions like these: What chance does the son of an unskilled worker have of attaining a professional or managerial position? What are the social origins of people in high-status occupations?

Table 6.1 is an "outflow" table that starts with fathers and asks about the mobility of their sons. For example, reading across the row that begins with "lower manual" on far left, you can trace the sons of fathers who held unskilled "lower manual jobs." Some of these sons (20%) rose to "upper–white collar" (professional or managerial) positions, but the largest group (36%) followed their fathers into unskilled manual jobs. The top row shows that the sons of upper–white collar fathers were much more successful or fortunate. Over 40% had jobs at the same high levels as their fathers; only a few (15%) sank to lower manual jobs. (See the notes at the bottom of the table for a fuller description of the five occupational categories.)

Several general conclusions can be drawn from this table:

1. There is a high level of occupational inheritance—sons following fathers into jobs at the very same occupational level.

Table 6.1   Outflow From Father's Occupation to Son's Occupation

Father's Occupation	Son's Occupation (in percent)					
	Upper–White Collar	Lower–White Collar	Upper Manual	Lower Manual	Farm	Total
Upper–White Collar	42	31	12	15	1	100
Lower–White Collar	34	33	13	19	1	100
Upper Manual	20	20	29	29	1	100
Lower Manual	20	22	20	36	2	100
Farm	16	18	19	35	12	100
Total (N = 3,398)	27	25	19	27	3	

Summary

Up	Stable	Down	Total
41	33	26	100

SOURCE: Author's analysis of General Social Survey data, 1995 to 2004.

NOTE: Occupational categories: *upper–white collar*—higher professionals and managers; *lower–white collar*—semiprofessionals, technicians, sales, clerical; upper manual—skilled blue collar; lower manual—operatives, service workers, laborers; farm—farmers and farm workers.

---

[2]These and similar tables in this chapter are based on our analysis of data from the periodic General Social Survey (GSS). Because mobility tables require a large number of cases to produce trustworthy results, we have aggregated data from several successive years of the GSS. We limit the data to people who are of prime working age (25–64) and in the labor force. "Father" as used here is a shorthand that includes, in a few cases, mothers, grandmothers, and other family heads.

2. The higher the father's occupation level, the better the son's chances for occupational achievement.

3. Nonetheless, there is also considerable movement up and down the occupational ladder from one generation to the next, as the summary statistics at the bottom of the table suggest.

4. By a considerable margin, sons are more likely to move up than down.

Table 6.2 presents the same data in an altered form that allows us to answer a different set of questions. It is an "inflow" table. Unlike the previous outflow table that started with fathers and traced the fortunes of their sons, this table starts with sons and asks about their fathers. Here we want to know where people at different levels in the occupational structure come from. To answer this question, the table is percentaged down the columns, rather than across the rows, like the outflow table.

Reading down the Farm and Lower Manual columns on the right side of Table 6.2, you will not be surprised to learn that farmers are typically sons of farmers and lower-manual workers are typically sons of lower-manual workers. But you may be surprised to learn from the statistics in the upper–white collar column that the people in these high-status occupations are of very diverse social origins. Although the majority are from white-collar backgrounds, many are from blue-collar or farm families. We get this result, despite the high level of father-to-son succession at the top, revealed in Table 6.1, because the relative number of professional and managerial jobs has been growing, creating opportunities for advancement from below.

The mixed social origins of people at the upper end of the occupational structure are relevant to a question we raise at various points in this book: what is the degree of consistency or clarity in the class system? Upward social mobility appears to reinforce consistency toward the bottom, while undermining it at the top. Many men in privileged occupational positions apparently grew up in relatively modest circumstances, and their lifestyles, values, and political outlook may, in some degree, differ from those who began life at the top of the pyramid.

Table 6.2   Inflow From Father's Occupation to Son's Occupation

Father's Occupation	Son's Occupation (in percent)					
	Upper–White Collar	Lower–White Collar	Upper Manual	Lower Manual	Farm	Total
Upper–White Collar	34	27	14	13	11	23
Lower–White Collar	23	24	13	14	11	19
Upper Manual	17	18	36	25	11	23
Lower Manual	19	23	28	36	19	26
Farm	6	7	10	14	49	10
Total (N = 3,398)	100	100	100	100	100	100

SOURCE: Author's analysis of General Social Survey data, 1995 to 2004.

NOTE: See Table 6.1 for category definitions.

Wealth Mobility

Most sociological studies of social mobility have focused on occupation rather than other stratification outcomes, such as income and wealth. They have done so because occupation is both important and fairly stable, from year to year. As a practical matter, occupation is a convenient variable because adults can more easily recall parental occupation than wealth or income level. But recently, longitudinal data have become available, making it possible to trace intergenerational income mobility and estimate intergenerational wealth mobility (Keister 2005; Mazumder 2005:93).

Table 6.3 examines the influence of parental wealth on the wealth of young adults, ages 35 to 43, in 2000. Parental households were ranked by their estimated wealth two decades earlier. Those who grew up in the wealthiest households have an obvious advantage. More than half of the adult sons and daughters of parents who ranked in the top fifth are themselves in the top fifth among young adults. Eighty percent are in the top two-fifths. Wealth mobility seems no more likely for those who grew up in the bottom fifth (first row). Most remain at or near the bottom. Between these extremes, the table reveals considerable movement, but usually up a step or down a step. As the summary figures at the bottom indicate, moving up is about as likely as moving down.

Table 6.3   Wealth Mobility: Outflow From Parents to Young Adults, in Percent

	Adult Son's or Daughter's Household Rank					
Parents' Household Rank	*Bottom Fifth*	*Second Fifth*	*Middle Fifth*	*Fourth Fifth*	*Top Fifth*	*Total*
Bottom Fifth	45	27	11	9	8	100
Second Fifth	24	35	20	14	7	100
Middle Fifth	11	20	35	21	13	100
Fourth Fifth	7	11	23	33	25	100
Top Fifth	5	6	9	25	55	100

Summary			
Up	**Stable**	**Down**	**Total**
31	41	28	100

SOURCE: Modified from Table 2.10 in Keister 2005:57.

NOTE: Adult children ranked by household net worth in 2000, from National Longitudinal Survey of Youth (NLSY). Parental household rank estimated from other family traits reported for NLSY in 1979.

Social Mobility of Women

Studies of social mobility have focused almost entirely on men. Until recently, this emphasis seemed justified by the limited labor force participation of women. But as we have seen, in the Age of Growing Inequality, families have grown increasingly dependent on women's paid labor.

Studying women's mobility presents a special set of problems. For one, with whom do we want to compare women workers—their mothers or their fathers? Many mothers in previous generations did not work or worked only part time or intermittently. They were not the economic mainstay or foundation of social status for their households. If the comparison is with fathers, we face the problem that women workers are distributed across occupations in a very different fashion than men, making father–daughter comparisons tricky. Women are, for example, more likely to be nurses, secretaries, sales clerks, and maids. If we see that many daughters of lower-manual workers have moved "up" to white-collar work, are we looking at a difference between generations or between genders?

Table 6.4, which compares the occupations of fathers and daughters, does not resolve these problems (a task beyond the scope of this book), but it does give us a preliminary picture of women's mobility. We have drawn on the same data we used to measure men's mobility but applied a different set of occupational categories to daughters, to capture the typical pattern of female employment. The one conclusion that we can confidently draw from this table is that women's occupational attainment, like men's, is powerfully influenced by class origins. For example, 54% of the daughters of upper–white collar men hold a broad range of managerial and professional jobs, including doctors and lawyers but also teachers and nurses.[3] In contrast, daughters of lower-manual workers are more likely to hold clerical, sales, technical, and service jobs, and a significant minority hold unskilled blue-collar jobs, most typically as factory workers. But there is also ample evidence of social mobility in the table. For example, a third of the lower-manual daughters have managerial or professional occupations.

Table 6.4   Outflow From Father's Occupation to Daughter's Occupation

| | Daughter's Occupation (in percent) | | | | | |
Father's Occupation	Prof. and Managerial	Sales and Clerical	Service	Blue Collar	Farm	Total
Upper–White Collar	54	33	9	3	1	100
Lower–White Collar	49	34	11	6	*	100
Upper Manual	35	37	18	8	1	100
Lower Manual	32	39	19	9	1	100
Farm	34	28	22	14	2	100
Total (N = 3,398)	40	35	16	8	1	

SOURCE: Author's analysis of General Social Survey data, 1995 to 2004.

NOTES: See Table 6.1 for father's category definitions. Daughter's professional/managerial is a broad category including many semiprofessional jobs. Sales/clerical includes technicians, most notably those in health fields. Blue collar includes a small proportion of skilled manual workers.

*Under .05%.

---

[3]The inclusion of these semiprofessionals makes the professional/managerial category used for daughters much broader than the upper–white collar category applied to fathers.

Circulation and Structural Mobility

We have been looking at the pattern of intergenerational movement up and down the occupational hierarchy. But what are the underlying causes of this mobility? To simplify the problem, imagine a completely closed society of male workers in which all sons replicate the positions of their fathers. Now consider the factors that could open it up. There are two basic ones.

1. *Circulation mobility.* Some sons might slip down the scale and thereby make room for others to climb up. In this instance, mobility is a "zero sum game"—that is, some must lose for others to gain.

2. *Structural mobility.* If technological and organizational changes occur in a way that creates jobs at a faster rate in the middle and upper levels of the occupational structure than in the lower levels, then some sons will have the chance to climb into the new positions without displacing anybody—not a "zero sum game" this time but a "win-win" situation. (Of course, occupational change could also reduce opportunities to move up and force some people down.)

In fact, both of these processes are at work in any modern society, though we have no practical way of determining precisely how much mobility to attribute to each. And we certainly cannot determine which is responsible for any individual's mobility (or lack of mobility). The world is more complex than our simple model suggests. For example, total mobility is inflated by what we might call the multiplier effect: If a new job is created near the top of the hierarchy, it is likely to be filled by someone from the middle, and that person's job in turn is taken by someone from below. Thus, one new opening can create two or more moves in a step-by-step progression.

We can, however, get a sense of the influence of occupational change by reexamining the occupational trend data we saw in Chapter 3. We saw that some occupational categories grew rapidly, while others expanded slowly or shrank. For example, from 1940 to 2000, the white-collar categories grew at several times the rate for the labor force as a whole, while the numbers of farm workers and laborers actually shrank (Table 6.5). The effect of these dual trends was to create what might be described as a mobility updraft, drawing people into higher positions. (The updraft was somewhat offset by the growth of service employments.)

During these decades, the categories of professional and technical, managers and proprietors, and clerical and sales taken together grew from about 10.5 million to 35 million positions—a 330% expansion. If these occupations had grown at the same rate as the average for the whole labor force, there would have been only 19 million men in white-collar jobs in 2000. The difference of the actual over this "expected" average growth was about 15 million positions: that many men had the chance to move up in the system to fill the newly created jobs. They constituted 21% of the male labor force in 2000—and this estimate is the absolute minimum because we have no way of adding those men who moved up because of the multiplier effect described earlier.

Table 6.5   Male Occupational Distributions, 1940 and 2000

	In Millions		
	1940	*2000*	*Percent Change*
Total	39.2	72.3	+84
Professional/technical	2.3	11.8	+413
Managers/proprietors	3.4	10.8	+218
Sales/clerical	4.8	12.1	+152
Craftsmen/foremen	6.1	13.5	+121
Operatives	7.1	2.6	+35
Service workers	2.4	7.2	+200
Laborers, except farm	4.7	4.4	−17
Farm occupations	8.5	2.7	−70

SOURCES: U.S. Census Bureau 1975; U.S. Department of Labor 2001a.

NOTE: Because of modifications in Census Bureau occupational categories and the shift from 14 to 16 as the minimum age for which occupational data are typically tabulated, these 1940 and 2000 distributions are not strictly compatible (see Table 3.4, especially 1972 tabulations). But they are an appropriate basis for the rough comparisons made here.

Declining Social Mobility

In earlier chapters, we examined trends in income, wealth, and occupational structure. What can we say about trends in social mobility? Is there more or less upward mobility than there was in the past? How has the pattern of mobility changed? To answer these questions, we compare the mobility of two groups of workers, the first interviewed in the 1970s and the second around 2000, and we single out younger workers, ages 25 to 44, whose experience most clearly reflects recent change. Note that the younger workers interviewed in the 1970s entered the labor force during the postwar Age of Shared Prosperity and those interviewed around 2000 began their careers in the current Age of Growing Inequality.

We find that upward mobility has, in fact, declined and downward mobility has increased in recent years. Among younger workers, upward mobility dropped from 45% to 37% (Table 6.6). The basic reason is that the mobility updraft created by the swelling of the higher occupational categories and the shrinkage of some of the lower categories had largely played itself out by 2000. There was, in other words, less structural mobility in later years. We confirm this with an Index of Structural Mobility, based on the difference between the occupational distributions of fathers and sons. The index shows a decline in structural mobility, especially for younger workers.[4]

Further scrutiny of the same data reveals that the sons of upper and lower white-collar workers are a little less likely to inherit their father's position. Their privileged background counts for less, though it still counts. This tendency indicates that the decline in structural mobility is being offset, albeit modestly, by an increase in circulation mobility.

---

[4]The index is the minimum percentage of sons who have moved because of changes in the occupational structure between generations. It is calculated by subtracting the percentage of fathers from the percentage of sons for each occupational category in which the latter exceeds the former. The sum of these differences is the index number.

Table 6.6   Trends in Social Mobility (in Percent)

	All Men		Men 25 to 44	
	1970s	2000	1970s	2000
Mobility				
Up	48	41	45	37
Stable	34	33	34	34
Downs	19	27	21	30
Index of Structural Mobility	18	11	16	8

SOURCE: Author's analysis of General Social Survey data, 1972–1979 and 1995–2004.

NOTE: Refers to occupational categories and sample as defined in earlier tables.

Conclusion

Our most consistent and predictable conclusion in this chapter is simple: in the race to the top, it helps to start there. This is confirmed by mobility tables, based on recent data, that reveal a strong association between the occupations of fathers and sons. Comparisons between fathers and daughters point to the same conclusion. Nonetheless, there is a great deal of intergenerational social mobility. Two-thirds of sons, for example, have moved up or down from their fathers' occupational level, and upward mobility is more common than downward mobility.

But when we sharpen our focus and compare the experience of younger men in recent years with an earlier cohort, we find evidence of stagnating opportunities. Although there is still net upward mobility, downward mobility is increasing. It is evident that the structural change that has fueled mobility in the past is waning. These findings regarding the experience of younger workers are our best window on the future. What they suggest is not encouraging.

These trends could have powerful political implications. Already many young workers fear that they will not have the same opportunities for advancement that their parents enjoyed and that they might not even be able to maintain the living standard that they grew up with. Especially in periods of economic stagnation, such feelings could turn them against politicians they identify with the status quo.

Key Terms Defined in the Glossary

circulation mobility

intergenerational mobility (see social mobility)

outflow mobility table/inflow mobility table

social mobility

structural mobility

Suggested Readings

Erikson, Robert and John Goldthorpe. 1992. *The Constant Flux: A Study of Class Mobility in Industrial Societies.* New York: Oxford University Press.
  Comparative mobility study covering the era of prosperity following World War II. American pattern seems quite similar to European.

Featherman, David and Robert M. Hauser. 1978. *Opportunity and Change.* New York: Academic.
  Detailed report on the last major national mobility survey (1973). Comparisons with the 1962 survey.

Hout, Michael. 1988. "More Universalism, Less Structural Mobility: The American Occupational Structure in the 1980s." *American Journal of Sociology* 93:1358–1400.
  Interpretation of mobility trends among men and women, 1972–1985.

Hout, Michael. 2004. "How Inequality May Affect Intergenerational Mobility." In *Social Inequality*, edited by Kathryn M. Neckerman. New York: Russell Sage Foundation.
  How mobility may (or may not) be related to inequality, with some observations on the history of mobility studies.

Keister, Lisa A. 2005. *Getting Rich: America's New Rich and How They Got That Way.* New York: Cambridge University Press.
  An examination of intergenerational wealth mobility.

Newman, Katherine S. 1988. *Falling From Grace: The Experience of Downward Mobility in the American Middle Class.* New York: Free Press.
  How ex-managers, laid-off skilled workers, unemployed professionals, divorced women, and their families respond to their fall from economic grace.

CHAPTER 7

Family, Education, and Career

The transmission of property from generation to generation, in the same name, raised up a distinct set of families, who, being privileged by law in the perpetuation of their wealth, were thus formed into a Patrician order, distinguishable by the splendor and luxury of their establishments. From this order, too, the king habitually selected his counselors of State. . . . To annul this privilege, and instead of an aristocracy of wealth, of more harm and danger than benefit to society, to make an opening for the aristocracy of virtue and talent, which nature has wisely provided for the direction of the interests of society, and scattered with equal hand through all its conditions, was deemed essential to a well-ordered republic.

Thomas Jefferson (1821)

n the quotation that begins this chapter, Thomas Jefferson, himself the wealthy son of a planter family, urges the replacement of the corrupt old "aristocracy of wealth" with a new "aristocracy of virtue and talent." The aristocracy of wealth, Jefferson thought, could be eliminated by changing the laws of inheritance so that men would divide their landed estates equally among all their children and thus eventually arrive at small holdings. The aristocracy of virtue and talent—characteristics Jefferson assumed were "scattered with equal hand" among all classes—could be cultivated by creating a public school system that would provide primary education for all citizens and then select the best students for further training in high schools and universities. In his later years, Jefferson founded the University of Virginia as the capstone to this system and was so proud of it, he ordered that his tombstone should record his two greatest accomplishments: the writing of the Declaration of Independence and the founding of the University of Virginia.

From Thomas Jefferson forward, American political leaders have endorsed high rates of social mobility. In doing so, they have reflected the general values held by most Americans, which accept some inequality in society, believing that people should get different rewards for different kinds of work, but also believing that each generation should start fresh and compete in a "fair" way for those rewards. In other words, Americans believe that young people ought to have careers based on their own talents and desires rather than have their lives determined by the class positions of their parents; they should have "equality of opportunity" but not necessarily "equality of result." Accepting the fact that some aspects of the talent and desires of the children are inherited from or shaped by the parents, Americans have believed, or at least hoped, that equality of opportunity can be achieved through the school system. If all children have access to good schools at all levels, regardless of the financial resources of their families, then the graduates should be able to compete on a reasonably fair basis.

The connections among family background, education, and career success, often the focus of heated debate, will be our main concern in this chapter. Here we shift away from the systemic or structural perspective of the last chapter to highlight the experience of individuals. Instead of asking how much mobility exists in the system as a whole, we ask why some individuals are more successful than others. Obviously, these are closely related topics. An individual's chances for success will depend on both the available opportunities and his or her personal characteristics, including family background and educational achievement.

In what follows, we are often interested in establishing the strength of the connections between variables—for example, the influence of father's occupation on son's occupation. We use two yardsticks for this purpose: simple correlation and variance explained. Both tell us how accurately we can predict the value of one variable by knowing the values of preceding variables. Correlation is measured by a coefficient that ranges from 0.0 (there is no relationship between two variables) to 1.00 (we can, with perfect accuracy, predict the second variable by knowing the first). In the social sciences, most correlation coefficients are intermediate. For example, the correlation between father's education and son's education is around 0.45.

Variance explained, which is closely related to correlation, is typically invoked when we are trying to measure the influence of a series of prior variables on a single dependent variable. Variance explained is expressed as a percentage of total variance in the dependent variable, with higher percentages expressing stronger relationships. We might, for example, determine that father's occupation and education together "explain" 45% of the variance in son's occupational success—meaning that the differences in occupation and education recorded among fathers statistically account for 45% of the differences in occupational level observed among sons. Other factors, both known and unknown, account for the remaining 55%.

Blau and Duncan: Analyzing Mobility Models

What makes the study of the social mobility of individuals complicated and debatable is this: everything is related to everything else. We know, for example, that a son's occupational attainment is significantly correlated with his father's occupational status and his own education (both approximately 0.50). These relationships are depicted in diagram A in Figure 7.1. But, as diagram B indicates, the father's occupational status is also correlated with the son's education. This introduces a complication. How are we to interpret the relationships among the three variables? One possibility, portrayed in diagram C, is a simple causal chain. Concretely, sons with high-status fathers have the opportunity to get a good education and, as a result, to get good jobs; sons with low-status fathers cannot get a good education and end up with crummy jobs. The effect of the father's socioeconomic status on the son's career is mediated through education.

But things might be more complicated—they often are. Imagine that a son with a high-status father completes his education and then gets an additional career boost: the old man hires his son or finds him a good job through a family connection. Diagram D adds another arrow to represent this possibility, but the additional arrow opens a new mystery. Which is more important, education or easy access to the job? Perhaps education did not really count for much and was just the inevitable result of a privileged background like, say, driving an expensive car; in the end, what really mattered was the job opportunity. Of course, the opposite is also possible or some unequal combination of the two influences.

Things would get even more complicated if we considered additional factors—for example, race and parents' education. A black child might get less encouragement in school than white schoolmates and end up quitting school early, only to face discrimination in the job market. This outcome would be less likely if the child's parents were well educated themselves. Of course, well-educated parents would be likely to have high-status jobs, an additional career advantage for the son or daughter. If we added new arrows to the diagram to represent these and other interrelated influences, we would soon have a tangle of causal pathways resembling a New York City subway map—a complicated image for the complex realities of social mobility.

In 1967, Peter Blau and Otis Dudley Duncan introduced an innovative methodology for analyzing the complexities of career success and failure. Their book, *The*

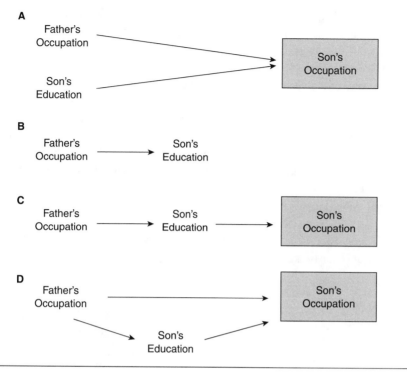

Figure 7.1   Causal Relationships in Occupational Attainment

American Occupational Structure, which analyzed the results of a large-scale national mobility survey, inspired much of the subsequent research in the field. Blau and Duncan depicted career development in formal "path diagrams"—more elaborate versions of our drawings—that became the basis for their statistical analysis of the survey data. Their goal was to sort out the influence of each variable, independent of the others. Path analysis, as this method is called, helped Blau and Duncan conceptualize two key aspects of the problem: chains of causation (father's occupation influences son's education, which in turn, shapes son's career prospects) and multiple causal pathways (father's occupation influences son's education and later directly influences his job search).[1] Note that path analysis emphasizes the temporal element in social mobility. Blau and Duncan's analysis dealt sequentially with family background, education, son's first job, and job at the time of the survey, examining the influences operating at each stage.

Figure 7.2 summarizes a later national mobility study that used the methods introduced by Blau and Duncan (Featherman 1979; Featherman and Hauser 1978). The study is based on a representative sample of adult men who answered questions about their family background, education, and career experience. Each respondent's occupation at the time of the survey was rated on a scale of socioeconomic status (SES) developed by Duncan and akin to the occupational rankings we

---

[1]Our examples are sons because the literature we are discussing here was limited to fathers and sons.

discussed in Chapter 2. The family background variables examined in the study include father's SES, father's education, parents' marital status, and race. The analysis sorts out the determinants of occupational attainment as measured by respondent's SES.

Our very simplified path diagram and the related pie graph in Figure 7.2 emphasize the study's main conclusions. The diagram indicates that both family background and education, as expected, contribute to son's socioeconomic status. It shows direct and indirect (through education) connections between family background and son's SES. The pie graph parcels out responsibility for the result (son's SES) between the background variables and educational achievement in terms of variance explained.

What do we learn here?

1. *Family background through education.* The main effect of family background (that is, the background variables taken together) is its influence on education,

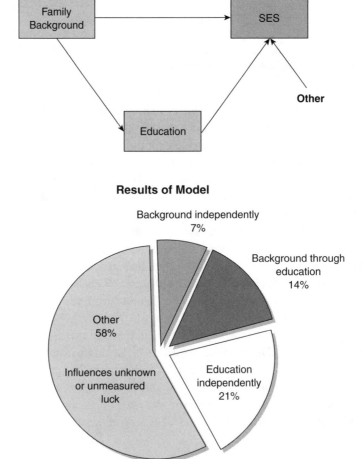

Figure 7.2   Factors in Social Mobility

which, in turn, contributes to career success (14% of total variance). In other words, rich kids generally get good educations; poor kids normally do not.

2. *Family background independently.* Family background also has a smaller, independent effect on son's SES (7%). This refers, for example, to the father who hires his own child or gets him into the union, as well as the negative effects of racial discrimination on blacks.

3. *Education independently.* Education has a substantial independent effect (21%). This tells us that educational achievement does more than just reflect the privileges or disadvantages of family background. Some rich kids flunk out and lose the career-enhancing benefits of education. Some not-so-rich kids graduate from college and get a big career boost.

The model suggests that total family background (1 and 2) and education independently (3) are equal in their influence on career success.

Together, they account for 42% of total variance. But this leaves 58% of the variance, corresponding to "other" in the diagram, "unexplained." As we will see later in this chapter, recent, more elaborate models reduce the amount of unexplained variance, but significant unexplained variance always remains.

There are two basic reasons for unexplained variance. One is that we cannot always measure our variables as precisely as we would like to. For example, the father's or son's occupation may be better or worse than it sounds from the rough occupational titles we use. The other is that there are many influences that we do not, and often cannot, include in the model. For example, some individuals are more ambitious and able than their academic accomplishments suggest. Personal charm or good looks may also contribute to career success, but they are not easy to measure.[2]

Some of the unexplained variance in mobility models is certainly the result of nothing more than dumb luck. Someone starts a career at just the right moment (the economy is booming, his field is hot) and gets ahead fast. Someone else experiences a traumatic event early in life and never quite gets over it. This is all very frustrating for sociology, but from a more poetic viewpoint, it's nice to know that life is not entirely predictable.

Jencks on Equality

Blau and Duncan inspired a large new literature on "status attainment," including two important books by Christopher Jencks and a research team at Harvard University. In the first book, Jencks and his colleagues (1972) attempted to integrate

---

[2]Unfortunately, many of the variables we might add turn out to be largely "redundant" because they overlap with variables already included. For example, this model does not include mother's education, but mother's education is highly correlated with father's education and occupation, which are included. Therefore, this additional variable would not contribute much to explained variance.

most of what was known into one grand model that would include more variables than Blau and Duncan had studied. The book was titled *Inequality: A Reassessment of the Effect of Family and Schooling in America*. It created quite a stir. Jencks did not collect original data but extended Blau and Duncan's results. He interpolated a lot of material from other studies and organized it into a series of discrete stages corresponding to points in the life cycle, instead of relying on a single path diagram that summed up the whole process. Jencks also added income as the final dependent variable in the causal chain, after SES. He found that there was only a modest correlation between occupation (SES) and income. This might seem surprising, but as we noted earlier, the occupational titles we use are fairly crude measures of what people do at work. Moreover, pay scales vary around the country, workers generally get higher pay as they accumulate seniority on the job, and unionization or the economic strength of a firm can have big effects on earnings.

As the Harvard researchers worked backwards in the causal chain, they found ever-weaker connections with income. A son's income correlated 0.44 with his SES, 0.35 with his education, and 0.29 with his father's SES. Every time we add another variable to a chain of causes and effects, we weaken the connection between the earliest predictors and the final outcome because at each step in the process, other factors are at work (or "luck" keeps reappearing). Consequently, this new model, which uses income as the final dependent variable instead of occupational prestige, reduced the measured impact of education on life chances.

The Harvard team also wanted to measure somewhat more sharply the "pure" effect of years of schooling, independent of the qualities that students bring with them to the school building. To do so, they added another variable to the model: the son's IQ at age 11, on the assumption that it is the best available measure of the talent or "cognitive ability" of the child that reacts with the stimuli of the school.

In organizing the vast statistical material from many different studies that is integrated in "inequality," Jencks tended to stress a polemic point. Economic (occupational and income) inequality in the United States is very large, and simply improving the quality of bad schools and reducing the differences among individuals in the number of years they attend school will not go very far in eliminating the economic differences. The reason is that education is not a strong determinant of income. In fact, Jencks estimated that barely a quarter of the variance in the incomes of adult men could be predicted by combining all the usual predictors: family background, IQ score, years of education, and even job title. In other words,

> Economic success seems to depend on varieties of luck and on-the-job competence that are only moderately related to family background, schooling, or scores on standardized tests. . . . The fact that we cannot equalize luck or competence does not mean that economic inequality is inevitable. Still less does it imply that we cannot eliminate what has traditionally been defined as poverty. It only implies that we must tackle these problems in a different way. Instead of trying to reduce people's capacity to gain a competitive advantage on one another, we would have to change the rules of the game so as to reduce the rewards of competitive success and the costs of failure. (Jencks et al. 1972:8)

If we really propose to make incomes more equal, Jencks is suggesting, we must do something that directly alters the operation of the labor market to reduce the range of incomes: corporation presidents should receive less, and people who fry hamburgers should receive more. The level of government intervention required to produce such a result is often called socialism, and Jencks was not afraid of adopting the word. He implied that our national preoccupation with changing the schools was a distraction from the real issue.

The reaction to Jencks's conclusions was strong; he was not faulted on technical grounds, but many critics were unhappy about the implications of the words luck and socialism. Jencks and colleagues set out on a second round of research, including the integration of several new surveys of careers (including those of brothers) that had since become available. The second book, *Who Gets Ahead?* (Jencks et al. 1979), was much less controversial, for at least two reasons. First, Jencks gave up the noble attempt of the earlier volume to communicate with the general public and instead offered dry and statistical prose in a highly technical style. Second, he avoided the discussion of policy implications for the most part. Jencks no longer emphasized the difficulty of creating more equality of result; he simply reported on the amount of variance in schooling, job, and income that could be explained by different background characteristics.

These policy weaknesses of the book were offset by a prodigious amount of new work, using many new sources of data, which certainly increased the accuracy of the calculations, although they did not dramatically alter the conclusions. Let us report a few of the more significant analyses.

Who Gets Ahead? made a number of changes in procedure, but the most interesting for our purposes comes from a recognition that the standard measures of family background, such as parents' education and occupation, are crude estimates of the total influence of social origins. Although Jencks provided those measures, he also added, where possible, a measure based on the similarity of the careers of brothers. Jencks assembled some new data of his own as well as material available in other studies, although the difficulty of tracing two brothers in each family keeps the samples much smaller than one would like. The resemblance between brothers is a total measure of family influence that is useful but not very specific: It covers shared genes, shared environment in the home and the neighborhood, and the possible influence of one brother on the other. It predicts career outcome somewhat better than conventional SES measures of family status, thus reducing the size of that bothersome residual variance.

Jencks and his colleagues also rounded up data on personal characteristics that went beyond the usual IQ scores, including measures of personality. They found that no single additional indicator was particularly powerful, but a combination of them predicted career outcome about as well as the IQ score by itself and seemed to be adding something new to the mix.

Let us start with the overall conclusion and then break it down into a series of sequences that follow one another. The resemblance between brothers, or total family background, explained almost half the variance in the occupational statuses of men and a little less than a third of the variance in their incomes. (If we could measure the lifetime earnings of men instead of using only 1 year as the dependent variable, the predictability would be even higher because temporary ups and downs would be averaged out.) Obviously, this result is much more powerful than Blau and Duncan's original model and suggests more inheritance of position than was previously believed.

The most important indicator of total family background was the father's occupational status, but that accounted for only a third of the resemblance between brothers. Adding in a string of other demographic variables (education of both father and mother, income of both, family size, and race) adds another third. The remaining third of the variance "is presumably due to unmeasured social, psychological, or genetic factors that vary within demographic groups" (Jencks et al. 1979:214).

If the schools are to be equalizers, then talented children of poor families must stay in school as long as talented children from rich families. Otherwise, the schools are just mediators or the transmission belts used by privileged families to obtain privileged careers for their children. If we measure talent at a relatively early age and then follow the subsequent paths taken by the children, we can estimate how much of school attainment is a consequence of talent (as measured by the tests), regardless of the socioeconomic background of the students' families. Jencks and his colleagues reported that early cognitive test scores explain about 40% of the variance in ultimate educational attainment; about two-thirds of that is independent of the connection between the IQ scores themselves and family background. This implies that a lot of variation in talent is picked up on the tests, even among families of similar social status (indeed, even among brothers), and the variation in talent has a significant amount of influence on educational attainment. However, test scores have far less effect on educational attainment than a completely meritocratic system would require, and they have even less effect, further out the causal chain, on earnings, predicting only about 5% to 10% of their variance (all else being equal). The schools are already more "fair" than the labor market, according to these particular measures.

In a general sense, everybody knows that staying in school pays off—that is, people with more education have higher-status jobs and earn more money. But how much difference does it make? And, is the difference linear, so that an extra year of high school is worth about the same as an extra year in college, or is the last year in college, which provides the coveted bachelor's degree, worth something extra?

Studies done after Jencks's first book was written indicate a slightly higher correlation between education and earnings than he first reported, probably more because of methodological improvements than because of actual trends in society. The various surveys indicated an average correlation of about 0.40 between years of education and earnings in dollars, without controlling for other influences.

The second book contained a chapter on education written by Michael Olneck (Jencks et al. 1979, Chapter 6), which shows that the relationship between years and dollars is not linear: Years in college are worth more than years in high school, particularly if one stays long enough to pick up a degree. Olneck found that years of high school bring an increase in dollar earnings over elementary school graduates of about 40%, and four years of college bring an increase of almost 50% over high school graduates.

Even when the results are controlled for family background and IQ, the high school diploma yields a 20% gain in lifetime income over elementary school completion and the college degree generates an additional 35% over the high school diploma. (The college figure is probably on the low side because the earnings gap between high school and college grads has grown significantly in the last two decades.) Thus, teachers appear to be correct when they tell students that staying in school is a good investment.

There appears to be a gradual reduction in the impact of background variables as one gets older. The payoff of higher education (especially for those who stay long enough to get the degree) is almost as great for those from poorer families as for those from more privileged backgrounds. Of course, the latter are much more likely to start life with the resources (especially economic and cultural capital) that will get them into and through college.[3] In the earlier years, then, education is as much a reflector of family background as an equalizer that offsets family background, but in the later years, education has more independent influence.

These measured effects of education fit with our discussion in Chapter 3, which emphasized a rather sharp break in the occupational system, dividing people with college degrees from all others. Many employers screen applicants through the simple device of academic credentials, and jobs at the managerial level, even for young people just starting to work who are being selected and trained for such jobs, are given to those with college degrees in hand. Since World War II, we have dramatically increased the level of education for all segments of the population: almost everybody now finishes elementary school, and more than 80% finish high school. Among those who never earn a college degree, there is a large pool of workers, and the jobs they get and the earnings they receive depend more on family background, individual talent, and the oddities of work experience than on a few years more or less of schooling.

But at the upper levels of the occupational system, the good jobs go primarily to those who have completed college—about a quarter of young men and women. Within that select group, the further differential impact of family background on jobs attained or dollars earned is rather small: it is the degree that counts. Of course, background has a lot to do with the chance of getting the degree in the first place, so we reach a double conclusion: College degrees both protect the privileges of people born into upper-status families and permit many from lower-status families to climb into the upper-middle class.

Who Goes to College?

Not all people with college degrees have outstanding careers, but few people (other than athletes and entertainers) now achieve important positions without them. And the income gap between the high school– and college-educated, as we have noted, is growing. Thus, the question posed is obviously an important one: who, exactly, goes to college?

---

[3]Lareau's work (2003), discussed in Chapter 5, is obviously relevant here.

William Sewell and his associates tackled this question in a massive study that followed the careers of 9,000 Wisconsin students who graduated from high school in 1957 (Sewell and Hauser 1975; Sewell and Shah 1977). They found that IQ, SES, and gender strongly and independently affected the graduates' chances of attending college. For example, a boy with below-average intelligence might go to college if he came from a high-status family. A boy from a low-status family was unlikely to continue his education unless he had a very high IQ. Girls, however smart they were, had slim prospects of attending college.

Some 25 years later, an even larger, well-financed national study produced remarkably similar results. The only big change was the virtual disappearance of the gender gap in college enrollment. The more recent study, titled "High School and Beyond" (HSB), was conducted in the 1980s. It focuses on the generation that includes the parents of many of today's college students. We analyzed part of the HSB data for this book.

Because gender was no longer important, we focused on the joint influence of social class and mental ability. The HSB researchers divided their national sample of high school seniors into four cognitive ability quartiles and four class or "socioeconomic status" levels. We have cross-tabulated the two variables in Table 7.1, which records the proportion of students in each IQ-SES subgroup who were attending college 2 years after graduation from high school.

The results at the extremes are wholly predictable: nearly all the high-ability graduates from high-status families, but very few low-ability graduates from low-status families went to college (83% vs. 13%). But what about kids in the middle, who are neither brilliant nor dim? How does class affect their prospects? Among *top SES high school grads with just below-average abilities* (second cognitive quartile), 57% were in college, compared to 33% of *bottom SES kids with just above-average abilities* (third cognitive quartile). The HSB data confirm the Wisconsin study findings that mental ability and social class are strong, independent determinants of educational attainment.

Table 7.1   College Attendance by Social Class and Cognitive Ability (Percent in College)

Cognitive Ability Quartile	Family Socioeconomic Status Quartile			
	Lowest	*2nd*	*3rd*	*Top*
Lowest	13	13	20	35
Second	23	24	40	57
Third	33	42	51	69
Top	51	63	74	83
Total (*N* = 11,995)	24	33	48	69

SOURCE: Author's analysis of "High School and Beyond" (HBS) data.

NOTE: Proportion of 1980 high school seniors attending college in 1982. Cognitive ability quartiles based on average of standardized scores on reading, math, and vocabulary tests.

Class disparities in education have proven quite resilient, despite efforts dating back to Jefferson's time to eliminate them. Figure 7.3, drawn from an analysis by Thomas Mortenson (2001), tracks college attendance rates of young adults by income quartile. The chart reveals enormous income gaps based on college attendance. In 2000, for example, 75% of 18- to 24-year-olds from families in the top income quartile (highest 25%) were currently enrolled in or had completed a year or more of college. For the bottom quartile, the figure was 35%. Although the proportion attending college has increased at all income levels since 1970, the income gaps have remained relatively constant.

Class disparities in actual completion of college are even greater. In 2000, more than 50% of 18- to 24-year-olds from families in the top quartile had college degrees, but only 10% of their peers from the bottom half of the income distribution had degrees. Moreover, this gap has grown much wider since the early 1970s (Mortenson 2001).

Class not only influences the chance of continuing education after high school, it also influences the type of school chosen and the chance of graduating. A substantial proportion of college students start at 2-year community colleges. These schools' curricula usually permit some students to transfer later to 4-year colleges. But many community college students will drop out of school before they complete 2 years, and others concentrate on 2-year technical courses that prepare them for positions that are somewhere in the middle of the occupational status range: computer programmers, dental assistants, automobile mechanics, and the like.

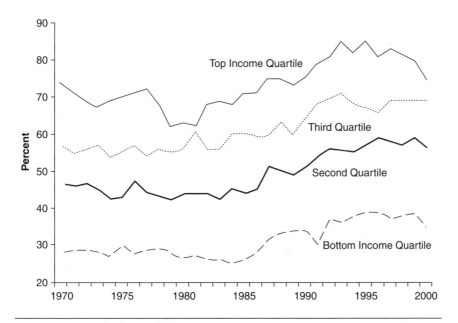

Figure 7.3   College Participation by Income Quartile, 1970–2000

SOURCE: Figure by Thomas G. Mortenson from *Postsecondary Education Opportunity*, October 2001. Reprinted by permission of Thomas G. Mortenson.

The whole system of higher education is stratified according to the quality of the education provided and the particular career preparation emphasized, and that hierarchy is paralleled by the stratification of students' families. Two-year community colleges tend to draw their students from the lower half of the income distribution. Private colleges and universities recruit disproportionate numbers of students from high-income families. The admissions profile of public 4-year colleges and universities places them somewhere in between.

The existing pattern of college attendance and graduation gives at least some support to two views of higher education in this country: (1) The American system of higher education is so big and so open that it provides major opportunities for talented youths from lower- and middle-class families to prepare themselves for successful careers that raise them above the level of their parents, and (2) the American system of higher education is sufficiently stratified that its main function is to reproduce for each generation of children the status positions held by their parents. Of course, both are true. The system is relatively open to the ambitious and talented, but it is also remarkably successful at reproducing the privilege of the privileged.

College and the Careers of Women and Minorities

In recent decades, class disparities in college degree attainment have, as we noted earlier, grown somewhat. However, gender disparities have reversed. More women are earning degrees than men. The black–white differential may have grown (Kane, cited in Neckerman 2004).

For those who do complete college, the degree is a valuable credential but not a guarantee of career success. In general, education yields greater economic returns to white males than to females or African Americans. Both blacks and women face discrimination in hiring and promotion. Women at all levels of education tend to be segregated into comparatively low-paying occupations. Women's earnings are further undercut by their tendency to move in and out of the labor force for family reasons. However, as we saw in Chapter 3, such differences have been narrowing.

By the 1990s, the average yearly earnings of college-educated African American men were over 70% of the earnings of their white counterparts—far from parity, but an improvement over 1949, when the ratio was just 52% (Jaynes and Williams 1989). One problem with this rough measure is that it mixes older black men, who began their careers when opportunities for well-educated blacks were very limited, with their younger counterparts, who encounter broader career horizons. A better way to evaluate the effect of education on the earning power of disadvantaged groups is to make such ratio comparisons between populations with carefully matched characteristics. For example, we can compare black and white men of the same age, education, and work pattern, as in Table 7.2.

The table, based on 2006 Census Bureau data, makes race and gender comparisons of median income for full-time, year-round workers, matched for age and years of education. People with exactly 4 years of high school or exactly 4 years of college are singled out, as are 25- to 34-years-olds. We focus on these younger workers because their careers should most reflect the changes promoted by the

Table 7.2   Earnings Equity Ratios, 2006 (in Percent)

| | Black/White | | Female/Male |
	Males	Females	All
All educational levels	78	90	76
High school	78	90	72
College	75	107	70
High school, 25–34	78	88	78
College, 25–34	83	98	80

NOTE: Based on median annual earnings of full-time year-round workers over 25.

civil rights revolution of the 1960s and the women's movement that began in the 1970s. Here are examples of how to read this slightly complicated table: The first figure in the bottom row tells us that a young (25–34) black man with a college degree earned 83% of what a similar white man made. The next figure across (98%) makes the same comparison between young black and white women. Finally, the last figure in this row tells us that young, college-educated women (of all races) earned 80% of what their male peers did. Of course, all of these refer to full-time workers.

Not evident from the table, which is based on relative comparisons, is the fact that education pays an absolute dividend for all categories of workers. For most demographic groups (females, white females, young black men, and so on), a college degree in 2004 was worth roughly $15,000 more in annual earnings than a high school diploma. For older white men, the gap is even larger.

We can draw two broad conclusions from Table 7.2: (1) Among younger workers (25–34), a college degree narrows racial and gender income disparities, and (2) the college payoff for black women is especially large. On average, college-educated black women earn 7% more than their white peers. These generalizations reinforce the idea that a college education pays off for disadvantaged groups and suggest that it may do so more in the future than it has in the past.

We can conclude that education—especially college education—is an equalizer, but with an obvious caveat: it is only an equalizer if people have good access to education. Today, access for women is at least equal to that for men, but access for blacks is well behind that of whites, and low-income young people have suffered a deterioration in their chances of getting a college degree.

Conclusion

The previous chapter looked at intergenerational social mobility from a macro or structural point of view. We focused on the gross pattern of mobility—how many people move up, down, or not at all. We considered the extent to which this movement reflected equality of mobility chances among individuals or simply opportunities generated by a changing occupational structure. In this chapter, we shifted perspective and concentrated on the determinants of individual success within the framework of gross mobility.

Research shows that two factors, family background and education, are strongly correlated with career success. But measuring their separate influence is complicated because the two are correlated with one another. In general, people from privileged backgrounds get a double boost for their careers: a direct advantage (for example, father's connections help his children get jobs) and an indirect advantage (they get more education). But education also exercises a strong influence that is independent of background. In effect, this means that the daughter or son in a blue-collar family who manages to get a college education is likely to do well. This conclusion is consistent with our finding in the last chapter that a remarkable percentage of the sons of manual workers move into upper–middle-class jobs.

Christopher Jencks, building on Blau and Duncan's earlier work, concluded that background variables (including father's occupation, parents' education, income, and race) account for nearly half the variance in occupational attainment. (This figure includes family influence on the extent of a son's education.) If Jencks's estimate is correct, we can make a good guess at a boy's odds of occupational success on the day he is born—a good guess, but not a sure thing: there is another 50% to be determined by factors ranging from personal initiative and charm to pure luck.

The literature on status attainment suggests that the influence of education on occupational success is of similar magnitude to the influence of family background. Income rises—this is hardly surprising—with years of education. However, Jencks's research revealed something less obvious: The final year of college, if it results in a degree, is worth much more than any of the preceding years. The power of the degree is such that, among those who manage to graduate, the influence of family background is greatly reduced.

But if college graduates are equalized in this sense, access to college remains very unequal. There is, for example, an enormous gap in the college participation rates of young adults from high- and low-income families. The gap has actually increased in recent years.

What would happen if recent trends were reversed—if there were a significant increase in the proportion of high school graduates who complete college? Would this reduce the influence of family background on career chances? It might. But it could also contribute to "degree inflation" similar to what happened with a high school diploma—once a valuable credential, now almost universal and therefore devalued. If a high percentage of the population earned bachelor's degrees, the best jobs would probably be reserved (as they increasingly are) for people with postgraduate degrees from prestige institutions, who would likely come disproportionately from privileged class backgrounds. The outcome, of course, would not simply depend on the supply of workers with education credentials, but also on the demand created by economic growth and the changing shape of the occupational structure. As we have seen, individual ambitions, abilities, and family advantages are only half of the mobility picture.

Key Terms Defined in the Glossary

chain of causation
correlation, simple
multiple causal pathways
path analysis

social mobility
social succession
socioeconomic status (SES)
variance explained

Suggested Readings

Arrow, Kenneth, Samuel Bowles, and Steven Durlauf, eds. 2000. *Meritocracy and Economic Inequality.* Princeton, NJ: Princeton University Press.
> *Important set of essays exploring relationships among IQ, schooling, occupation, and income.*

Bowles, Samuel, Herbert Gintis, and Melissa Osborne Groves, eds. 2005. *Unequal Chances: Family Background and Economic Success.* New York: Russell Sage Foundation.
> *Important collection of research papers on the influence of family background on earnings and income.*

Lemann, Nicholas. 2000. *The Big Test: The Secret History of the American Meritocracy.* New York: Farrar, Straus & Giroux.
> *The history of the SAT. Created to break the grip of a social elite on the Ivies, the test provided the basis for a new elite, as privileged and exclusive as the old.*

Lucas, Samuel Roundfield. 1999. *Tracking Inequality: Stratification and Mobility in American High Schools.* New York: Teachers College Press.
> *Formal tracking based on discrete ability groups is a thing of the past. But informal tracking persists, to the disadvantage of lower-class children.*

McLeod, Jay. 2004. *Ain't No Makin' It: Leveled Aspirations in a Low-Income Neighborhood.* Revised ed. Boulder, CO: Westview.
> *Engaging portrait of black and white teens in a public housing project. Shows how their occupational aspirations are shaped by peer group, family, and school.*

Stevens, Mitchell. 2007. *Creating a Class: College Admissions and the Education of Elites.* Cambridge, MA: Harvard University Press.
> *How selective institutions reproduce class privilege: First hand study of the admissions process at a liberal arts college.*

CHAPTER 8

Elites, the Capitalist Class, and Political Power

Those who hold and those who are without property have ever formed distinct interests in society.

James Madison (1787)

[Campaign finance has become] an elaborate influence peddling scheme by which both parties conspire to stay in office by selling the country to the highest bidder.

Senator John McCain (2000)

James Madison and Alexander Hamilton were political opponents. But these two signers of the Constitution agreed on this much: politics and social class are unavoidably linked. Why? Because there is an inevitable conflict between: those whom Madison, above, calls "the propertied and the propertyless" and Hamilton described as "the few and the many."[1]

Chapters 8 and 9 examine the connection between politics and social class. We see these chapters as complementary. In Chapter 8, we deal with power focusing on "the few," those at the top of the class structure and the apex of the political order. In Chapter 9, we take up class consciousness, concentrating on "the many, the mass of the people."

Our emphasis on the few in this chapter is typical in studies of power and might appear to be guided by the inherent logic of the subject. After all, those at the top have the best opportunity to exercise power. But "the many" are far from powerless, especially—and here is where class consciousness comes in—when they are united by a sense of common identity and shared interests.

Three Perspectives on Power

We may define power as the potential of individuals or groups to carry out their will even over the opposition of others. Here we focus on issues of national power, beginning with an examination of three competing theoretical perspectives: elite, class, and pluralist. The elite perspective makes a sharp distinction between an organized minority (the elite) that rules and an unorganized majority that is ruled. Elite theories often focus on specific institutional bases of power. The class perspective, which has its origins in Marxist theory, also focuses on a ruling minority, but class theory is more specific about the identity of the rulers and the structure that creates them: they are the owners of productive wealth, or the capitalist class. The pluralist perspective denies that power is concentrated in one group. It maintains that in democratic societies, there are multiple bases of power representing the interests of competing groups, and no minority can easily impose its will. Obviously, the first two approaches have much more in common with one another than with the third.

A final prefatory note: To avoid improper usage and conceptual confusion, we consistently use the term *elite* as a collective noun like class or jury. Such terms refer to groups rather than individuals. In this chapter, elite (singular) alludes to some notable group (business executives) and elites (plural) connotes two or more such groups (executives, military officers, and public officials). These terms should not be used to refer to individuals, as in, "three elites went to a movie."

Mills: The National Power Elite

We begin our discussion with C. Wright Mills's classic work, *The Power Elite* (1956). Critical of American institutions at a time of growing domestic prosperity and (like

---

[1] See Hamilton's observations in the epigraph at the beginning of Chapter 1.

our own) of perceived international threat, *The Power Elite* inevitably evoked controversy when it was published in the 1950s. Mills characterized this self-satisfied era in our national life as "a material boom, a nationalist celebration, a political vacuum" (p. 326). His description of national political arrangements suggested that the growing power of a few was undermining American democracy. Although *The Power Elite* is over 50 years old, it is worth examining in detail because the issues it raises are still relevant for contemporary students of national power.

Mills's conception of the national power structure centered on the growing significance of three major interlocking institutions: the modern corporation, the executive branch of the federal government, and the military establishment. He saw each of these institutions becoming enlarged and centralized. A few hundred major corporations were taking the place of thousands of smaller competing firms that had once typified the economy.

The federal executive had gathered enormous powers and resources previously nonexistent or scattered among other units of government. The military, once small and decentralized, had developed into a colossal bureaucracy, commanding a war machine of unprecedented scale and destructive power. As the corporations, the federal government, and the military grew, they eclipsed and subordinated other institutions:

> No family is as directly powerful in national affairs as any major corporation; no church is as directly powerful in the external biographies of young men in America today as the military establishments; no college is as powerful in the shaping of momentous events as the National Security Council. Religious, educational, and family institutions are not autonomous centers of national power; on the contrary, these decentralized areas are increasingly shaped by the big three in which developments of decisive and immediate consequence now occur. (Mills 1956:6)

The implication Mills drew from these trends was that the basis of national power had been reduced to control over the three key institutions. Those who sit at the "commanding heights" of the corporate, political, and military hierarchies make the critical national decisions. Who are they? In the corporations, an amalgam of very rich families with corporate-based fortunes, plus the ranking executives of the top national firms, whom Mills collectively labeled the "corporate rich"; in the federal government, the president, vice president, cabinet, heads of major agencies, and members of the White House staff; among the military, the generals and admirals. These three institutional elites together constitute Mills' *power elite.*

Mills argued that the emergence of a national elite undercut important traditional bases of power. Community elites decline in importance as the power of national institutions grows. Investment decisions that are crucial for a community may be made in a distant corporate boardroom. Even the significance of personal wealth as a power resource is reduced. Very sizable individual fortunes seem lightweight relative to the massive assets of any major national corporation. However, the largest family fortunes, such as those of the Rockefellers, Mellons, or Fords, are typically invested in national corporations, and in this form, personal wealth can retain its significance as a basis of power.

Mills conceived the national structure of power as consisting of three tiers (see Figure 8.1). The top tier is, of course, the power elite. The bottom tier, which Mills labels "mass society," encompasses the great majority of the population. Subject to large-scale national institutions beyond their control or comprehension and misinformed by media that are dominated by national elites, the members of the mass are passive participants in the political system. Between the mass society and the power elite are "the middle levels of power," comprising a multitude of competing interest groups from labor unions to the gun lobby, whose typical arena of conflict is the Congress. The middle levels of power are the source of most political news, but they are not, according to Mills, the locus of the most important political decisions. In the welter of competing interests, none can impose itself. This "semi-organized stalemate," as Mills characterized it, only reinforces the dominance of the power elite. The key decisions are, of course, reserved for the power elite. But, which are the key decisions? Mills is unambiguous on this point. Two issue areas are of sweeping importance: economic policy and national security. These matters unmistakably, often brutally, intrude into the lives of ordinary men and women as boom or bust in the economy and peace or war abroad.

In Mills's schema of national power, his conception of the military as a separate, more or less autonomous elite evoked the most skepticism, even among those generally sympathetic to his argument. Mills seems to have mistaken the militaristic direction of foreign policy in the Cold War era for the power of the military over policy decisions. But since George Washington led the revolutionary army, the American military has been subject to civilian elites, who are responsible for the major war/peace decisions that concerned Mills. This was demonstrated most recently in the Bush administration's fatal decision to go to war in Iraq.

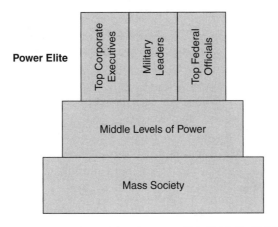

Figure 8.1   Mills's Conception of National Power

Mills, His Critics, and the Problem of Elite Cohesion

Mills's pluralist critics accused him of assuming what needs to be proven. If his elite was as powerful as Mills claimed, it should be able to impose its policy preferences in national decision making. But, pluralists noted, Mills had not tested the power of the elite by examining actual decisions. In this section, we focus on a related issue that has provoked extensive debate among Mills's readers: the problem of elite cohesion—that is, the extent to which the members of a hypothesized elite hang together in pursuit of common objectives and in opposition to other groups.[2]

The cohesion issue was effectively posed in a key essay by Robert Dahl, directed at Mills and other elites theorists, whom Dahl (1967) accused of "confusing a ruling elite with a group that has a high potential for control" (p. 28). To be politically effective, potential for control must be coupled with "potential for unity." The American military has the potential to impose a dictatorship on the nation, but that potential means nothing unless military leaders agree on that objective. Mills defined an elite in terms of key positions in organizations that possess vast resources (potential for control). But he failed, according to Dahl, to demonstrate a political consensus among the members of his elite (potential for unity).

Mills's pluralist critics are clearly predisposed to the belief that unity among power contenders is difficult to achieve, especially in contemporary America. This view found expression in David Riesman's *The Lonely Crowd* (1953), a pessimistic but influential book on American society and culture. Riesman asked two questions about power in America: "Is there a ruling class left?" and "Who has the power?" His answers were, respectively, no and no one. As the first question implies, Riesman believed that the country had a ruling class in the past. Early in the history of the republic, the ruling class consisted of the landed gentry and mercantile interests that constituted the Federalist leadership and, later, of captains of industry. But by the 1950s, the ruling class had been supplanted by an amorphous constellation of "veto groups"—organized representatives of specialized interests that included "business groups, large and small, the movie-censoring groups, the farm groups and the labor and professional groups, the major ethnic and major regional groups" (p. 246). The veto groups are distinguished from the powerful of previous eras by their inability to take positive initiatives to impose their own will. Feeling themselves powerless, chary of offending other groups, their function is largely defensive, "to neutralize those who might attack them" (p. 247).

How can any decision be made in such a political context? Is anyone in charge? Riesman's reply was that leadership may be needed to initiate something new or halt something in progress, but little leadership is needed to maintain the status quo. To the extent that anyone exercises power, it is over very specific and narrow issues. Power that might be effective over a broad range of issues or in the face of big questions that affect the nation as a whole is smothered by the action of the veto groups.

---

[2]For other lines of criticism, see Dahl 1961, 1967; Domhoff and Ballard 1968; Polsby 1970.

Mills (1956) characterized Riesman's amorphous power structure as "a recognizable although a confused statement of the middle levels of power, especially as revealed in congressional districts and in Congress itself" (p. 244). Power at that level is indeed a semi-organized stalemate. In Mills's view, Riesman was guilty of a mindless empiricism that equated all interest groups and all issues. Banks and organizations representing motorcycle riders are both concerned with national legislation. But to describe them as two veto groups misses the point. Some groups are more important than others because they have powerful resources at their command and because they deal with issues that are more vital to the nation. As we have seen, Mills was only interested in the big economic and national security questions and regarded as trivial most of the other issues that consume the attention of the middle levels of power.

But cohesion presented a particular problem for Mills. He had to demonstrate both that the three distinct elites are internally cohesive and that they are drawn together into a single power elite. Mills did present a series of mechanisms through which elite unity might be achieved. They fall into two categories: social-psychological and structural. The social-psychological mechanisms include similarities in origins, education, career, and lifestyles, which produce "a similar social type" and contribute to ease in informal association (Mills 1956:19). Mills presented evidence on these topics for each elite. He noted that elite men tend to be drawn from upper-class or upper–middle-class, urban, white, Protestant families and that they are likely to be educated in Ivy League schools. He found a significant overlap between the world of the power elite and upper-class "society," with its elaborate links among "proper" families, select prep schools, distinctive class values, and notions of style. (These commonalities, as Mills conceded, apply more to the civilian elites than to the military.)

In addition, members of the three elites have similar career experiences, even if they do not move through the same institutions. The corporate rich, the "political directorate," and the chiefs of the Pentagon share the experience of managing large organizations. The character of modern bureaucratic life has tended to blur the distinction between leadership in a large corporation, a civilian department of government, and an army. Mills contended that these shared elements of background and careers and the considerable material rewards attached to elite position tend to make members of the power elite conscious of the differences between themselves and the great mass of the population and to draw them together; they develop a form of upper-class consciousness, which leads them to view the world from a similar perspective.

The structural mechanisms of cohesion examined by Mills concern the more-or-less formal connections between institutions. One critical link is the interchange of personnel among the three institutions, especially the movement of representatives of the corporate world into and out of top political positions. Another tie between these two is the dependence of political candidates on financing from the corporate rich. The military is closely allied with the corporations, who are its suppliers, while the militaristic foreign policy pursued by the political directorate strengthens its ties to the generals and admirals. All three elites are, of course, compelled to consider each other by virtue of the inevitable interdependence of institutions operating on such a scale.

In sum, the pluralists are persuasive when they argue that Mills must prove that his power elite is cohesive. Otherwise, it is little more than a list of important people. In reply, Mills points to a series of factors that tend to unify the individual members of the power elite and draw the three major institutional sectors together: common social background, shared lifestyle and values, similar job experience, interchange of personnel, campaign financing, and institutional interdependence. In this chapter, we examine contemporary evidence on most of these topics.

Power Elite or Ruling Class?

If pluralists believe that Mills fails in his efforts to prove that his tripartite elite is cohesive, one Marxist critic contends that Mills is all too successful. In a perceptive essay on *The Power Elite*, Paul Sweezy contends that there is an unresolved tension in the book between two views of the elite. The first is based on social class: Mills provided evidence that "those who occupy the command posts do so as representatives or agents of a national ruling class which trains them, shapes their thought patterns, and selects them for their positions of high responsibility" (Sweezy 1968:123).

Much of the evidence Mills presented for elite unity seems to point in this direction—for example, his emphasis on the higher class origins of members of the power elite and the recruitment of capitalist-class leaders to key cabinet positions. The second view Sweezy finds in *The Power Elite* focuses on the bureaucratic elites at the top of three "major institutional orders"; here Mills treated the corporate, military, and political realms as distinctly separate domains with autonomous leadership, which come together to form the power elite. Sweezy was highly skeptical of the second view, particularly given the evidence that Mills presented for the first. The American military, Sweezy contended, is firmly under civilian control, and the political elite is dependent on the class that rules the corporations; thus, the justification for thinking in terms of three discrete institutional elites collapses.

Sweezy's criticism suggests an alternative to both the Millsian and the pluralist views of national power: the identification of power with the class that controls income-producing wealth, which he calls the ruling class. This approach—the last of the three theoretical conceptions of power we mentioned at the beginning of the chapter—proposes that Mills's "corporate rich" have largely subordinated competing elites to their will. Mills (1968) never dealt at length with this line of criticism, although in a breezy reply to critics on the left, he commented, "They want to believe that the corporation and the state are identical. . . . I don't believe it's quite that simple" (p. 224). It probably is not "quite that simple." But Mills appears to underestimate the power of wealth because he wants to fit it into a broader conception of elite power.

Who Rules?

The issues raised by *The Power Elite* have been reexamined in contemporary works by Dye (2002); Lerner, Nagai, and Rothman (1996); and Domhoff (2006), which we describe in this section. Dye's and Domhoff's contributions, it should be noted, are revised versions of books originally published around 1970.

According to Thomas Dye (2002), America is an "elitist" society. Citing writers from Alexander Hamilton to the Italian political theorist Gaetano Mosca, Dye asserts at the very beginning of his book that elite rule is inevitable in *all* societies, from the simplest to the most advanced. In particular, he stresses that a society such as the United States, which is dependent on large institutions, is ruled by those who hold the top institutional positions—the elite. Here he sounds like Mills, though his conception of the elite is much broader.

Much of Dye's book, *Who's Running America?* (2002), is devoted to defining the top institutional positions and describing the people who hold them. His elite consists of those who hold 7,314 leadership positions in 10 key sectors of American society. He emphasizes the concentration of the resources under their control:

> Individuals in these positions control more than one-half of the nation's industrial and financial assets, over half of the assets of private foundations and two-thirds of the assets of private universities; they control the television networks, influential newspapers and media empires; they control the most prestigious civic and cultural organizations; they direct the activities of the executive, legislative and judicial branches of the national government. (P. 139)[3]

Dye focuses on the phenomenon of "interlocking directorates." He notes that the 7,314 positions of elite power are held by only 5,778 individuals. In other words, some members of the elite hold multiple positions, sometimes five or more. These "interlockers" connect 32% of the positions. Corporations seem especially likely to be connected in this fashion, both with other corporations and nonbusiness institutions. From their vantage point, the interlockers are able to take broader view of common problems and contribute to elite consensus.

With his sprawling elite, Dye is compelled to address the problem of cohesion. He points to interlocking directorates and other bonding mechanisms, which, by and large, will be familiar to readers of Mills. Most members of the institutional elite, he finds, share upper–middle-class origins, though 30% are from upper-class families. (The current relevance of these figures is questionable, however, since they dates from the early 1970s [Dye 1976:152–153; Dye 2002:151].) Dye puts particular emphasis on the foundations, "think tanks," and policy planning groups that are discussed later in this chapter, as venues where the elite can explore problems and develop shared solutions. Although he describes internal factions within the elite, he seems to regard it as relatively cohesive.

In *American Elites* (1996), Lerner, Nagai, and Rothman are skeptical of claims for elite cohesion. They argue for a pluralist conception of national power based on differentiated elites. Note that the authors, unlike Mills or Dye, refer to elites in the plural. They see a division of national power among 12 "strategic elites" with distinct functional responsibilities. The leaders of business, the media, and the

---

[3]It is, unfortunately, impossible to be sure of the time period Dye's data refer to. Sometimes he suggests 1980, sometimes 1990.

military are examples. This brand of pluralism does not deny that power is concentrated in the hands of a few, but denies that elites exercise much influence outside their own sectors. There is, in other words, no core elite like Mills's power elite.

American Elites is based on surveys conducted in the 1980s. Probably because it is difficult to obtain interviews with CEOs, four-star generals, and the like, the researchers adopted a broad conception of elite status. The corporate respondents, for example, were from "upper and middle management." The government officials were "high-ranking bureaucrats" not appointed by the president. The interviews focused on the social backgrounds and opinions of the members of the 12 elites, topics which are directly relevant to the question of elite cohesion. By showing that strategic elites differ in their origins and viewpoints, they expected to undermine the notion of a unified central elite.

A chapter titled "Room at the Top" describes a pattern of recruitment to elite positions that favors the children of privilege, but not exclusively. The great majority of respondents report that their families had average (33%) or above-average (38%) incomes when they were children. A little more than half had fathers who were professionals or managers, well above the national proportion of men in these categories during the years these leaders were growing up. (The researchers do not differentiate respondents of upper-class origin.) But many had fathers with less impressive jobs and below-average incomes. Almost all their elite respondents completed college, but surprisingly few (31%) graduated from highly selective institutions.[4] Most members of the strategic elites appear to be from upper–middle-class backgrounds, but there is enough diversity among them to doubt that shared background could be a critical basis of elite cohesion.

The authors claim to have found a high-level ideological division among American strategic elites. But the evidence they present is not decisive. Asked to classify themselves as conservative, moderate, or liberal, respondents differed in predictable ways: corporate executives and military officers generally took the conservative label; majorities of labor leaders, movie makers, journalists, bureaucrats, and religious leaders described themselves as liberals. But answers to specific opinion questions reveal a more complicated picture. For example, responses to a series of questions about the desirability of liberal social and environmental policies show the bureaucrats quite close to the conservative thinking of business and the military. More liberal responses to these items came, as expected, from labor officials and public interest group leaders. Some questions revealed an unexpected degree of inter-elite consensus—agreement, for example, that the legal system favors the wealthy. Virtually no one thought that corporations should be publicly owned, and strong majorities in the various elites (except labor and public interest) agreed that business is "fair to workers." The right to abortion was strongly supported by all but the religious elite. In short, *American Elites* does not document a clear pattern of diverging opinion among the strategic elites.

---

[4]Detailed tabulations in the book show that these generalizations hold up fairly well for the 12 elites taken individually. The biggest exceptions were labor leaders, who are, as might be expected, largely from blue-collar backgrounds, and corporate lawyers and public interest group leaders, who come disproportionately from higher class backgrounds.

G. William Domhoff (2006) in *Who Rules America?* presents an interpretation of power in America that he describes as "a class theory." But he begins by examining an elite, the directors of major corporations. (A corporation's board of directors, as Domhoff explains, is its legal governing body, typically composed of officers of the company and so-called outside directors). Using recent (2004) data on the directors of almost 2,000 large corporations, Domhoff found that they linked firms in a loosely connected network he labels the corporate community. The average firm had 6.1 interlocks with others.

Domhoff's analysis also revealed an elite within the elite—15% to 20% of directors who sit on multiple corporate boards. These individuals tend to be associated with the largest corporations. They are also likely to be on the boards of nonprofit organizations, to participate in business leadership organizations, and to assume government positions. They are, concludes Domhoff, the "inner circle" of the corporate community.

The next step in Domhoff's (2006) argument is his claim that the corporate community is "closely intertwined with the upper class" (p. 49). Here his evidence is staler and weaker. Domhoff is referring to the upper class in a social rather than economic sense. In a 1960s version of this book, Domhoff (1967) had concluded that 53% of corporate directors could be considered upper class, based on social indicators (p. 51). Now he cites Dye's finding that 30% of the corporate elite is of upper-class origins, but this number is from Dye's (1976) early-1970s research (pp. 152–153). Domhoff notes that corporate directors are likely to belong to elite social clubs like the Links Club in New York and the Pacific Union Club in San Francisco. Here again, the evidence is from the 1970s. That aside, it is not clear whether membership in such clubs is simply recognition of the prestige of corporate position or an indication of broader participation in an upper-class status community. Domhoff's own recent research shows that the corporate community has been diversifying to include more minorities and women. But he writes that their presence in the boardroom does not change the atmosphere, since they tend to come from the same class and educational background as their white, male counterparts.

The third piece in Domhoff's theory is corporate participation, through funding and membership on relevant boards, in the same foundations and policy organizations that also interest Dye. Although the information he presents is dated, it is likely that these ties continue to be significant since such organizations need corporate financing and their work is of continuing interest to the corporate elite. Domhoff illustrates the relationship among the corporate community, the social upper class, and the policy-formation organizations with a diagram consisting of three modestly overlapping circles. Within this universe of partially overlapping groups, he defines what he calls "the power elite" as those who sit on the boards of corporations, the boards of corporate-controlled policy organizations, or both. Members of the upper class are only part of this power elite if they fall into one of these two categories. The power elite, suggests Domhoff (2006), "provides a leadership basis for the exercise of power on behalf of the owners of all large income-producing properties . . . those who have a stake in maintaining the current wealth and income distributions" (p. 103).

Although they define the structure of power differently, Dye, Domhoff, and the authors of *American Elites* all accept Mills's premise that power in this country is concentrated in the elites at the head of large organizations in key sectors of American society. All find that the members of their elites are from relatively privileged backgrounds, though they differ on the details. All deal, in one way or another, with the problem of elite cohesion. Dye finds a more or less unified elite of 5,000 leaders (a remarkably small number) of the dominant organizations in 12 key sectors. *The American Elites* authors describe multiple elites (plural) whose power is limited to their own sectors and whose divergent views limit collaboration. Domhoff finds one dominant elite, the leaders of an interconnected corporate community, linked to a social upper class and the network of policy organizations.

The National Capitalist Class: Economic Basis

We devote the remainder of this chapter to an examination of the capitalist class in its economic, social, and political aspects. This discussion touches on many of the issues that came up as we discussed elite theories.

We have defined the capitalist class as consisting of people who receive most of their income from invested wealth. Although such people exist at virtually all income levels, only at very high levels does dependence on property income become the predominant pattern. A traditional division within this class is that between local and national capitalists. The national capitalists are those who own or manage major national corporations; Mills collectively dubbed them "the corporate rich." The locals are affluent but community-oriented business people, such as local media owners, real estate investors, and large local retailers. In recent decades, this distinction has become less meaningful. Heirs to large local fortunes have tended to convert them into diversified national wealth. The community banks, newspapers, and other enterprises owned by their families are often sold to national corporations.

In Chapter 4, we found that households with the highest incomes derive most of their income from accumulated wealth, in such capitalist forms as interest, dividends, and rents or from lucrative family-owned businesses (see Table 4.3). We discovered that wealth—especially corporate wealth—is highly concentrated in the United States. Even after the stock market boom of the 1990s, the majority of American households owned no corporate stock or mutual fund shares (Cashell 2000:7). The richest 10% of households owns almost 80% of corporate stock. Within this stratum of wealth-holders, the top 1% of all households holds close to 40% of corporate stock, 60% of equity in private businesses, and close to half of investment real estate equity (Table 4.7).

Our data on the top 1% come from government surveys. But this source tells us relatively little about the largest American fortunes at the top of the capitalist class. For that purpose, we turn to the list compiled annually by *Forbes* magazine of the 400 wealthiest individuals in the United States. The net worth of the 400 individuals listed by *Forbes* in 2006 ranged from $1 billion to $53 billion. As recently as 1991, the list ran from $275 million to $5.9 billion. Table 8.1 contains a sampling of the *Forbes* 400, with information on the size and source of their fortunes. The list begins with Microsoft founder Bill Gates (*Forbes* 1991–2006).

Table 8.1   Large Fortunes

Personal Fortunes	Estimated Net Worth (Billions)	Primary Source of Wealth
William Gates III	53.0	Microsoft
Steven Ballmer	13.6	Software
Barbara Cox Anthony	12.6	Newspapers, Radio, TV
Jacqueline Mars	10.5	Candy
John Kluge	9.1	Metromedia
Philip Knight	7.9	Shoes
Edward Crosby Johnson III	7.5	Mutual Funds
Donald Newhouse	7.3	Publishing
Samuel Newhouse, Jr.	7.3	Publishing
Michael Bloomberg	5.3	Financial News
Micky Arison	5.0	Carnival Cruises
Lester Crown	4.1	Investments
Ralph Lauren	3.9	Fashion
Richard DeVos	3.5	Amway
Paul Milstein	3.5	Real Estate, Banking
Joan Tisch	3.4	Loews
Edgar Bronfman, Sr.	3.2	Seagram Co.
Donald J. Trump	2.9	Real Estate
Riley P. Bechtel	2.7	Engineering
Amos Hostetter, Jr.	2.6	Cable Television
Henry Kravis	2.6	Leveraged Buyouts
David Rockefeller, Sr.	2.6	Inheritance (oil)
Perry Bass	2.5	Oil, Investments
Gordon Getty	2.3	Inheritance (Getty Oil)
Carl Lindner	2.3	Insurance, Investments
Herb Allen	2.0	Investment Banking
Irwin M. Jacobs	1.7	Qualcomm
James France	1.6	Banking
Thomas Siebel	1.5	Siebel Systems
Arthur Blank	1.3	Home Depot
John Sperling	1.3	Apollo Group
Hope Van Beuren	1.3	Inheritance
Roy Disney	1.2	Walt Disney
David Duffield	1.2	Peoplesoft
Robert Fisher	1.2	Gap
Jerome Kohlberg	1.2	LBOs, Investments
James Clark	1.1	Netscape
Frederick A. Krehbiel	1.1	Molex
Alexander Spanos	1.1	Construction
Julian Robertson, Jr.	1.0	Money Management

SOURCE: *Forbes*, Oct. 9, 2006.

Forbes touts the 400 list as evidence of America's open class structure. In recent years, the list has been invaded by younger, often technology-oriented, entrepreneurs. According to *Forbes*, about half the people on the list are, like Gates, "self-made millionaires." An independent analysis of the 1997 *Forbes* list paints a more complicated picture, with four levels of inherited wealth:

1. *Inherited 400 Status*—42% of the *Forbes* 400 inherited wealth worthy of a place on the list. Examples are David Rockefeller, grandson of the founder of Standard Oil, and the Newhouse brothers, who inherited a large newspaper-magazine empire, including such properties as *Vogue* magazine.

2. *Inherited Significant Wealth*—13% of listees inherited more than $1 million or a substantial enterprise, or received equivalent start-up capital from a family member. Edward C. Johnson, for example, inherited Fidelity investments; Forest Mars, Sr., inherited his parents' candy business.

3. *Inherited Lesser Wealth or Advantage*—14% were from affluent or socially upper-class backgrounds, probably below the $1 million level. Bill Gates fits here. His father was a successful lawyer. Gates attended a private school.

4. *No Inherited Advantage*—31% began their careers with no apparent financial or social advantage. For example, John W. Kluge, an immigrant raised by his mother in a Detroit tenement, built a media empire after World War II (Collins and Yeskel 2000:64–67).

Obviously, many of the superrich got their fortunes the old-fashioned way: they inherited it. Ironically, Steve Forbes, the publisher of the magazine and sometime presidential candidate, is himself in this category.[5] But 45% (categories 3 and 4) of the listees accumulated 400-level fortunes more recently, beginning with modest advantages or none at all. This suggests a new wave of fortune building, comparable to developments a century earlier.

By focusing on individual net worth, the *Forbes* 400 and similar listings underestimate the concentration of wealth at the top. Many of the largest fortunes are held in common by members of extended families. These clans are based on descent from a founding ancestor, who typically accumulated the initial fortune during the rapid expansion of American capitalism in the late-nineteenth and early-twentieth centuries. Some are represented among the current 400. But in many cases, shares in these established family fortunes are sufficiently dispersed after two or three generations that no individual controls the $1 billion now needed to earn a place on the *Forbes* list.

Forbes has sometimes published lists of large family fortunes. A 1998 article listed 50 families, all with fortunes in excess of $1 billion (Forbes 1998). Among those included were the Grahams, publishers of the *Washington Post;* the Coors brewery family; and the Duponts, who control a large share of the chemical company. This short list did not reach down to families with fortunes in the hundreds of millions, like the Sulzbergers, who control the *New York Times,* and the politically prominent Kennedys, whose collective fortune *Forbes* estimated at $350 million in 1991.

---

[5]Forbes probably belongs on the list himself, but the magazine has never included its publisher/owner.

The concentration of personal wealth is paralleled by the concentration of wealth in the corporations themselves. The dominant position of the largest corporations seems to have been further consolidated since Mills wrote. In 1950, shortly before the publication of *The Power Elite*, the 100 largest U.S. industrial corporations (among nearly 200,000) already controlled approximately 40% of all industrial assets; by the 1990s, their share had grown to 75% (Dye 1995:15, 19).

Mills observed that even the largest personal fortunes were small, relative to the assets concentrated in large corporations. Economic power, he contended, belonged to those who controlled the corporations. But who controls the corporations? Mills's answer, as we have seen, was "the corporate rich": top corporate executives and extremely wealthy families like the Fords with substantial stakes in major corporations.

But as major corporations have grown larger, their stock has become more dispersed, and their relationship to even the largest stockholders has grown more distant. The biggest corporations have hundreds of thousands of stockholders; individual holdings of even 5% (enough to give the owner significant influence over management) are relatively rare. At the same time, many wealthy families have sought to reduce the risks to their fortunes by spreading investments across different corporations and sectors of the economy. Today, owners are rarely seen at helms of major corporations.

Small and medium-sized corporations are still likely to be controlled by their owners—typically families or small groups of stockholders, including many of the *Forbes* 400 listees.[6] But by 1980, only 22 of the 100 largest industrial corporations were controlled by owners. Most are run by professional managers, generally free of owner influence (Herman 1981:61). The current exceptions to this rule tend to be relatively young corporations, like Jeffrey Bezos's Amazon or Lawrence Ellison's Oracle, that have experienced spectacular growth in new sectors of the economy.

The displacement of owners at the top of major corporations set the stage for the spectacular rise in executive compensation of recent years. In 2006, a typical CEO of a major corporation earned $15.2 million—in real terms, six times the average of 1980. At the same time that executive compensation has reached unprecedented heights, it has become more dependent on the financial performance of the corporation. Incentives, including yearly bonuses and long-term stock purchase plans (called stock options), are added to base pay to reward executives who enrich their stockholders. For senior corporate officers just below the rank of CEO, the pattern is similar: high and rapidly climbing rewards, with total compensation more dependent on return to investors (DeCarlo 2007; *New York Times* 2001; Useem 1996:243–250).

Although top executives seldom hold even modest shares of the billions of dollars of outstanding stock in corporations they run, they often accumulate millions of dollars of stock and stock options in their companies. For this reason, we did not

---

[6]*Forbes* also publishes an annual list of the 500 largest privately owned (not publicly traded) companies. Most have few stockholders. Included are some large corporations such as Levi Strauss, Gallo Winery, and Fidelity Investments. In 2001, all had revenues in access of $600 million. Many were owned by individuals included in the *Forbes* 400 list (*Forbes*, 2001b).

hesitate to include them in our capitalist class, along with the owner-managers of smaller corporations.

The National Capitalist Class: Social Basis

Parallel to the economic basis for a national capitalist class in the corporate economy, there is a social basis in an upper-class social world built on prestige and exclusive patterns of association. Among the institutions identified with this world are the select prep school, the *Social Register,* and the elite metropolitan social clubs. These three have been widely used by researchers as formal indicators of membership in a socially defined upper class.

We have already described the appearance of the *Social Register* in early industrial America, when the new rich were being socially merged with the established upper classes. Although occasionally capricious in its inclusions and exclusions, a century after its creation, the *Social Register* remains, as Mills (1956) once described it "the only list of registered families . . . the nearest thing to an official status center that this country, with no aristocratic past, no court society, no truly capital city, possesses" (p. 57).

A small circle of prestigious prep schools, such as St. Paul's (New Hampshire), Hotchkiss (Connecticut), Foxcroft (Virginia), and Chapin (New York), traditionally draw most of their students from upper-class families. These day schools and boarding schools tend to be concentrated in the Northeast, but they draw many students from throughout the country. They are secular or nominally Episcopalian (the religious affiliation most common in the upper class) and traditionally single-sex institutions, although many have become coeducational in recent years. A few are older than the republic, but most were founded or experienced their major expansion around the time the *Social Register* appeared. They have, moreover, traditionally served a similar function: the integration of old prestige and new money (Baltzell 1958:292–319; Domhoff 1970:9–32).

Prep school graduates are informally referred to as "preppies," a term with mixed undertones of admiration and derision, which is used more loosely to refer to various aspects of an upper-class lifestyle (Birnbach 1980). The extension of the term is not inappropriate. The style and values the prep schools inculcate in their students equip them for participation in an upper-class social community. The network of personal ties that develop among prep school students and their families will serve them well in their subsequent careers and social lives. Prep schools contribute to a pattern of upper-class endogamy by bringing students and their siblings into contact with potential marriage partners, both directly and through upper-class social functions to which prep school students are likely to be invited. Among the latter are the debutante balls, the traditional events at which young women of the upper class are presented to society.

Most major metropolitan areas have one or two elite social clubs, such as New York's Knickerbocker, San Francisco's Pacific Union, or Philadelphia's Philadelphia Club, with generally upper-class memberships. The clubs provide an informal setting where upper-class associations can be developed and maintained and, on occasion, important business or political matters can be discussed free from outside scrutiny (Baltzell 1958:336-354; Domhoff 1970, 1974).

The prep schools and elite metropolitan clubs draw not only from their own regions but from a broader upper-class population. In this sense, they perform a national integrating function. It is not surprising that George W. Bush's upper-class family sent him from Texas, where he grew up, to study at Andover in Massachusetts, or that years later his father used San Francisco's exclusive Bohemian Club to introduce him to a group of friends who might be useful to the son's own presidential ambitions (Domhoff 2006:58).

As we saw in earlier chapters, social scientists have long emphasized the affinity between the worlds of wealth and prestige. Weber noted that the rich tend to draw together into upper-class "status communities," with common lifestyles, values, and patterns of association. Just such a social upper class developed around the wealthy "X family" in newly industrialized Middletown, as described by the Lynds (see Chapter 3). On a national level, industrialization brought a social merger of the traditional upper class and the owners of new industrial fortunes. By 1940, virtually all the founders of great fortunes in the late nineteenth or early twentieth centuries had traceable descendants listed in the *Social Register*.

Studies in the 1960s and 1970s pointed to the continuing link between national wealth and prestige. Domhoff (1967) and Dye (1976), in research reviewed earlier in this chapter, traced the upper-class connections of corporate leaders. For a study of "society" weddings, Blumberg and Paul (1975) tabulated occupational data on the fathers of brides and grooms; nearly 60% of the men whose occupations they identified were corporate executives.

For the student of national power arrangements, the significance of a link between upper-class society and corporate wealth is related to the problem of cohesion raised earlier. The achievement of consensus on specific policy issues and, more generally, the maintenance of class solidarity are made easier when those who own and control the major concentrations of national wealth encounter each other in a private sphere of informal relations. The schools and clubs are merely the outer manifestations of this realm whose deeper meaning resides in shared experience, intimacy, and the bonds of friendship and kinship that produce a consciousness of common identity and common values.

Domhoff (1974) points to group dynamics research in social psychology, which has established that physical proximity among the members of a group, frequent contact, a group reputation of high prestige, and an informal atmosphere all contribute to group solidarity. These are characteristic features of the upper-class world we have been describing and prepare us for another basic conclusion of the same research: "Members of socially cohesive groups are more open to the opinions of other members and more likely to change their views to those of other members" (pp. 89–90, 96). E. Digby Baltzell (1958), in his earlier study of upper-class Philadelphia, made a related point, which he phrased in terms of social control: An upper-class community inculcates and sustains "a mutually understood code of conduct" in its members. Upper-class people are especially subject to the "norms and sanctions of their peers. A man caught in an act of dishonesty or disloyalty fears, above all, the criticism of his class of lifelong friends" (p. 61).

In the 1950s, when Mills and Baltzell were writing, and as late as the 1970s, it was still possible to speak of the Protestant (or WASP)[7] "Establishment": an informal network of wealthy, powerful men who (1) were drawn from the upper-class social world described here; (2) held many of the key positions in industry, finance, and law; and (3) often served in high government office and, in or out of office, influenced national policy, especially economic and foreign policy. Never a formal organization, much less an elite conspiracy, the Establishment was more a set of personal relationships among people bound by social background, common values, and shared experiences.

The social world that produced the Establishment has not disappeared. The traditional prep schools are thriving, young women are still presented at debutante balls, the exclusive social clubs still function in major cities, and the *Social Register* is still published. The people who grow up in this privileged world certainly have much better than average chances for successful careers. But there is no longer an Establishment, in the sense we have described.

The Establishment faded away in part because recruitment to positions of power in business and government has become more open and meritocratic. In the past, graduates of exclusive prep schools could expect more or less automatic admission to Ivy League colleges, leading to top positions in business, finance, or law. But in the 1960s, the Ivies and other selective colleges broke this key link between social position and career success when they began to depend on the new SAT exam and began to give more emphasis to academic potential than family background in admissions. The change worked to the advantage of ambitious middle-class and upper–middle class students with public school educations.[8]

At the same time, both colleges and businesses came under pressure to eliminate discriminatory practices that severely restricted the access of those who did not fit the establishment's WASP-male profile to top positions. Today, women, Jews, Catholics, and blacks, though still underrepresented, are much more likely to serve on corporate boards, in the president's cabinet, and in other elite positions than they were in the 1960s.[9]

Economic change also contributed to the demise of the Establishment. There are fewer family-dominated companies, like the former Chase Manhattan Bank, long led by David Rockefeller, a quintessential Establishment figure. On the other hand, the postindustrial economy has given rise to large new fortunes in areas such as information technology, media, and communications, much as the new industrial economy did a century ago. These developments raise interesting, unanswered questions. Will the new money merge socially with old money, as it has in the past, to produce a renewed upper class? Will the social upper class become more diverse

---

[7]WASP, connoting White Anglo-Saxon Protestant, but meaning something like adult "preppie."

[8]It remained possible for very wealthy families to buy places for their sons and daughters at selective institutions through donations to university endowments. But this mechanism did not require upper-class credentials, just money.

[9]See D. Brooks 2000; Davidson, Pyle, and Reyes 1995; Judis 1991; Lemann 2000; Zweigenhaft and Domhoff 1998.

in religious and ethnic terms, reflecting the diversification of national elites?[10] We cannot even begin to answer such questions with the existing, dated literature on upper-class society. Perhaps future research will satisfy our curiosity.

For now, we can say that there is significant overlap—though probably a good deal less than in the past—between the upper-class social world and the national capitalist class and that the social institutions of the upper class provide some of the glue that binds the members of the capitalist class together.

The National Capitalist Class: Participation in Government

In December 1960, at a time when the Establishment was still very much alive, President-elect John F. Kennedy was selecting his cabinet. One of the people he turned to was a quintessential Establishment figure, Robert Lovett, a Wall Street investment banker with impeccable social credentials and extensive corporate connections (Halberstam 1972:16). In a perceptive book on the Kennedy-Johnson years, journalist David Halberstam relates their conversation, in which Kennedy artfully flattered Lovett, a Republican who had voted against him, and offered him a major cabinet post.

> Lovett declined regretfully . . . explaining that he had been ill. . . . Again Kennedy complained about his lack of knowledge of the right people, but Lovett told him not to worry, he and his friends would supply him with lists. Take Treasury, for instance—there Kennedy would want a man of national reputation, a skilled professional, well known and respected by the banking houses. There were Henry Alexander at Morgan, and Jack McCloy at Chase [Manhattan Bank], and Gene Black at the World Bank. Doug Dillon too. Lovett said he didn't know their politics . . . (their real politics of course being business). At State, Kennedy wanted someone who would reassure European governments: They discussed names, and Lovett pushed . . . young fellow Dean Rusk over at [the] Rockefeller [Foundation]. He handled himself very well, said Lovett. The atmosphere was not unlike a college faculty, but Rusk had stayed above it, handled the various cliques very well. A very sound man. (Pp. 16–17)

The three top cabinet appointments made by Kennedy, a liberal Democrat, reflected the advice of Lovett and other Establishment conservatives like him. For secretary of the treasury, he chose C. Douglas Dillon, an investment banker connected with Dillon, Read, and Company, a major Wall Street firm started by Dillon's father; for secretary of defense, Robert McNamara, who had just been made president of Ford Motor Company; and for secretary of state, Dean Rusk, then

---

[10]Graham (1999:51, 59–61) notes that some members of the black upper class now participate in traditionally white debutante balls and attend traditional WASP boarding schools.

president of the Rockefeller Foundation, a man with many admirers in corporate circles (Burch 1980:175–177).

Cabinet recruitment studies show that Kennedy's appointments followed a pattern inherited from his predecessors and maintained by his successors. Although the Establishment and figures like Lovell are phenomena of the past, presidents continue to appoint the kind of sound, reassuring men he recommended to Kennedy. Contemporary cabinets are more ethnically diverse and more likely to include women, but cabinet officers are still overwhelmingly drawn from the top of the class structure, most notably from the national capitalist class and the national prestige class we have been describing.

Beth Mintz (1975) and Burch (1980) examined the social and occupational backgrounds of people who served in the cabinet between 1897 and 1980. Their work shows that at least two-thirds of cabinet officers have been corporate officers, investment bankers, or corporate lawyers, often with clear upper-class social connections. The prominence of investment bankers and corporate lawyers among cabinet officers is notable; these two groups have long played a crucial mediating role between business and government, akin to the coordinating roles they play within the corporate world. Our own, more focused survey of the people who have held the top cabinet posts (State, Defense, and Treasury), from Kennedy's inaugural cabinet in 1961 through Bush's in 2001, produced similar results: 65% of these cabinet officers were drawn from major corporations, financial institutions, or corporate law firms.[11]

If business is well represented among top federal decision makers, there is no parallel representation of labor. Since 1913, only six men have served in the cabinet who were in any way connected with the labor movement. Most served for short terms and were secretaries of labor (Burch 1980:377; *Statesman's Yearbook* 1979–1990).

Ironically, modern presidents have generally come from lower levels in the class structure than have their own top cabinet secretaries. Many grew up in modest circumstances and made careers in politics, rather than accumulating fortunes in business. Lyndon Johnson, Richard Nixon, and Bill Clinton fit this mold. Two recent exceptions are George H. W. Bush and his son George W. Bush, descended from an affluent, socially prominent Connecticut family. Both made considerable fortunes on their own before entering politics.

What can be said of the class background of Congress? Evidence going back to 1906 indicates that most members of Congress come from business or the professions. They have typically been lawyers, executives, or small business owners (LTV Corporation 1990; Nagle 1977). According to the financial disclosure forms that members of Congress file annually, at least two-thirds of senators and 40% of House members are worth a million dollars or more, a status they share with the

---

[11]For this purpose, we have drawn on Brunner 2001; Dye 1995:81–85; *Who's Who in America* 1980; and *Who's Who in American Politics* 1979.

top few percent of American households. Almost 20% of senators and 10% of House members are worth $10 million or more.[12] The late Senator Daniel Patrick Moynihan was close to the truth when he observed, "We've become a plutocracy. . . . The Senate was meant to represent the states; instead it represents the interests of a class" (*New York Times*, Nov. 25, 1984).

Congress, then, is recruited from the upper levels of the class structure, although often from strata a notch or two below that of the top cabinet officers. Members of the House and Senate are also more likely to be career politicians rather than top executives or corporate lawyers on temporary assignment, as are many cabinet officers. The law and business backgrounds of many members of Congress suggest a smaller scale, more localized version of the corporate world so amply represented in the cabinet. If the cabinet is recruited from the national capitalist class, the Congress draws on local upper–middle and capitalist classes.

What do cabinet and congressional recruitment patterns tell us about national politics? It certainly cannot be argued that class position allows us to predict the political behavior of individual decision makers. For example, Edward Kennedy, a liberal Democrat, and John Warner, a conservative Republican, who seldom vote together, are among the wealthier members of the Senate. But we have already seen that the behavior and opinions of people in the aggregate are shaped in important ways by class; in the next chapter, we will see that this generalization can be applied to politics. Moreover, we have noted how informal association shapes receptivity to opinions. The senator or representative whose personal associations are largely capitalist class or upper-middle class is likely to be more open to viewpoints common at the top of the class structure than to those prevalent toward the bottom.

The precise effect of these tendencies is not easy to gauge, but we can point to certain issues, such as health care, that are of much greater concern to working-class people than to people of higher class rank. The failure of the United States to develop a national health care system may be related to the recruitment patterns of cabinets and congresses. On the other hand, when Senator Diane Feinstein, a relatively liberal California Democrat, voted in 2001 to eliminate the estate tax, she was doing her heirs a big favor. Feinstein, who grew up in a wealthy family, is worth tens of millions, according to her financial disclosure statement. Like many senators, she moves in an affluent social milieu, where she is likely to encounter people who regard the estate tax repeal as a vital issue (Graetz and Shapiro 2005:105).

---

[12]Estimate based on data from 2005 financial disclosure filings provided to us by the Center for Responsive Politics (CRP). The filings by senators and representatives exclude the value of their residence(s), the major asset held by most Americans. Filers specify a range (e.g., $15,000 to $50,000) for each asset or liability. Summing the item-by-item range limits, CRP calculated a minimum and maximum net worth for each member of Congress. We took the midpoint between these two figures as the member's estimated net worth.

Money and Politics

- Included in a bill that passed the Senate in late 1996 was a provision of particular interest to Frederick Smith, founder and chairman of Federal Express Corporation. Smith, whose fortune would climb to $2.2 billion by 2006 (*Forbes* 2006), personally lobbied senators for the provision, which makes it nearly impossible for FedEx workers to form unions and gives the company a distinct advantage over its unionized competitor, UPS. Federal Express has long been generous to lawmakers and their campaign treasuries. In 2006, campaign donations from company affiliated sources totaled almost $2 million.[13]

- Legislation signed by President Bush in 2001 placed the federal estate tax on a gradual path to extinction in 2010 (though the fate of the tax after that year was left in doubt). The campaign to repeal the inheritance tax, which affects only the top 2% of estates, was promoted by the Tax and Policy Group, a low-profile organization amply funded by several dozen national capitalist-class families, including heirs to the Mars (candy), Gallo (wine), and Campbell (soup) fortunes. With virtually limitless funding available to them, the proponents of repeal hired some of Washington's best lobbyists, pollsters, and public relations people and financed sympathetic policy research. Among their backers were some of the most generous contributors to political campaigns (Graetz and Shapiro 2005).

- In July 2007, the federal minimum wage, much eroded by inflation, was raised, for the first time in over a decade. The increase was resisted by business interests, among them some generous campaign contributors. Like the 1996 law mandating the last increase, the 2007 legislation was sweetened for opponents with several billion dollars in business tax relief.

These three cases illustrate the power and limits of political money in the hands of wealthy individuals and organized business interests. Labor unions lobbied hard against the Federal Express provision, but were overwhelmed by the company's well-financed efforts. The capitalist-class proponents of estate tax repeal hired Washington's best political talent to help them transform the way Congress and the public view what they relabeled "the death tax." With meager resources, their opponents could not respond in kind.[14] In contrast, business interests failed to block the minimum wage increase, a generally popular measure, once the Congress returned to Democratic control in 2007. But they retained sufficient influence to extract compensatory tax relief for themselves.

Money is not the only significant political resource. But it is the one resource that the rich have in abundance. This fact has long excited popular suspicion and encouraged periodic attempts to regulate political money. In the wake of the

---

[13]All figures on campaign donations cited in this section, unless otherwise noted, are from the Center for Responsive Politics Web site (www.opensecrets.org).

[14]The legislation did not, however, quite settle the issue, which must be contested again, sometime before 2011.

Watergate scandals, which drove Richard Nixon from office in 1974, Congress passed campaign funding reform legislation. The Nixon campaign had solicited enormous contributions from corporations; accepted illegal donations; and, in at least one instance, changed federal policy in exchange for a campaign donation.

The Watergate-era reform laws imposed strict limits on individual campaign donations to federal candidates and parties; regulated group donations, which must now be made through registered Political Action Committees (PACs); provided for federal financing of presidential campaigns on a matching basis for candidates who accept limits on campaign expenditures; and required public disclosure of campaign donations and expenditures. The legislation prohibits corporations and labor unions from contributing money from their treasuries to political candidates. But they can use their resources to run PACs that solicit donations from union members, corporate executives, or stockholders.

Perhaps the most significant benefit of the new campaign finance laws was transparency: We now have a better idea of who is contributing, how much they are giving, and how the money is being spent. Initially, the major effect of the reform legislation was to check the influence of very rich individuals on candidates, while increasing the political weight of certain organized groups, especially corporations. In the 1972 campaign that produced the Watergate scandals, some donors had given hundreds of thousands of dollars to candidates for federal office. Direct contributions to candidates on that scale became illegal under the 1974 reforms. As a result, the wealthy contributor became less important than the corporate PAC.

Currently, an individual contributor can give an aggregate total of approximately $100,000 to all federal candidates, PACs, and party organizations during a 2-year election cycle. But the contributor can only give a few thousand dollars to any single candidate. These amounts are high from a middle-class perspective, but modest enough to compel political fund-raisers to broaden their class focus. Congressional candidates who still depend on individual contributors for the greater part of their campaign money typically spend long hours on the phone and at fund-raising events chatting with the relatively affluent. They cannot depend on a few extremely wealthy people (Corrado et al. 2005; Malbin 2006).

Who makes campaign contributions? Studies going back to the 1920s indicate that contributors are, not surprisingly, better educated, higher in occupational status, and richer than the average American. A 1997 study of donors to congressional campaigns revealed that 81% had incomes over $100,000 and almost half had incomes in excess of $250,000, which would place them in the top 1% of households (Brown et al. 1995:7; Kevin Phillips 2002:328; Sorauf 1992:3, 34). Even as candidates learned that they could raise large sums in small contributions ($50, $250) via the Internet, their campaigns remained heavily dependent on the larger donations raised by more traditional methods.

The 1974 legislation, though strengthened by the 2002 McCain-Feingold law, was not wholly successful in eliminating very large donations of the magnitude associated with the Watergate scandals. Over time, interpretation by the courts, weak enforcement, and aggressive testing of legal limits by candidates and parties weakened the effect of campaign finance laws. Individuals cannot make large contributions directly to candidates or parties, but can make limitless donations to so-called "527s," politically motivated organizations named for a provision of the tax code, that are presumed to operate independently and do not explicitly endorse candidates. (In fact, they are only semi-independent and the campaign ads they sponsor typically support candidates or attack their opponents without quite saying vote for or against candidate X.) During the 2004 election campaign, 113 individuals donated $250,000 or more to such groups, including 25 who gave at over $2 million each. Most of the latter were on the *Forbes* list of the wealthiest Americans (CRP Web site; Weissman and Hassan 2006).

Courts, moreover, have ruled that no limit can be placed on the amount that candidates or their families can spend on their own campaigns.[15] Self-financing is especially important for nonincumbents. Many spend hundreds of thousands, even millions of dollars. The impulse is understandable. Many U.S. House and Senate elections are no contest at all, because the incumbent has raised many times more than the challenger. Although millionaires who finance their own campaigns typically lose, some wealthy individuals appear to have bought their way into the U.S. Senate, including Jay Rockefeller, Herb Kohl, Bill Frist, Maria Cantwell, and Jon Corzine. Wall Street millionaire Corzine invested $63 million in his successful Senate campaign. After Corzine left the Congress to become governor of New Jersey, Frank Lautenberg spent $1.5 million of his own money to win the same Senate seat.

Most campaign money comes from business sources, such as individual corporate executives and corporate PACS. As Table 8.2 indicates, business contributions far exceed labor contributions and tend to favor the Republicans. They account for nearly three-quarters of the approximately $2 billion in regulated campaign contributions covered by the table.[16] Another $424 million—more than half of

Table 8.2   National Campaign Contributions, 2004

	Millions of Dollars			Percent	
	Individuals	*PAC*	*Total*	*Rep.*	*Dem.*
Affiliation					
Business	1,267.8	240.3	1,508.1	55	45
Labor	1.3	60.1	61.4	13	87
Ideological	21.1	52.2	73.3	51	49
Others/Unknown	384.4	1.1	386.0	46	54
TOTAL	1,674.6	354.2	2,028.8		

SOURCE: Data from Center for Responsive Politics Web site www.opensecrets.org.

NOTE: Ideological includes proponents of gun control, right-to-life, and other political causes.

---

[15]The only exception is for presidential candidates, who voluntarily accept the limits that come with public financing—something major party candidates are no longer doing.

[16]The business proportion of individual contributions may, however, be somewhat overstated in the table because of the way donors are classified by the Center for Responsive Politics, the source of these campaign statistics.

which came from individual contributors who gave $100,000 or more—was given to "527" organizations (Weissman and Hassan 2006:81, 92).

The campaign finance landscape shifts from election to election with changing fund-raising strategies, jurisprudence, regulations, and legislation. What has remained constant is the predominance of the affluent and business interests in campaign giving. To the degree that efforts to regulate campaign finance have been successful, they have reduced—by contribution limits and disclosure requirements—the capacity of individual donors to buy political favors for themselves or their companies. Contributors are thus more likely to exercise influence by joining with others who share their class interests or ideological preferences.

Business Lobbies

We have examined the political influence of the capitalist class exercised through direct participation in government and through campaign finance.

We now turn to a third channel of influence: directly persuading public officials to adopt (or abandon) particular policies. Especially since the emergence of the modern corporate economy, business representatives have actively lobbied the Congress and the federal departments and regulatory agencies that carry legislation into practice. In the late nineteenth and early twentieth centuries, corporate lobbies—backed by abundant flows of cash—moved Congress with dependable ease. For example, a lobbyist for the National Association of Manufacturers (NAM) publicly acknowledged that he had bought legislative favors with bribes and influenced House leaders to appoint congressmen favorable to NAM to House committees and subcommittees (*Congressional Quarterly* 1976:654, 662). In recent decades, major business lobbies have generally employed more subtle methods, partly because competing interests are better organized and the possibilities of unfavorable publicity are greater.

Currently, the principal business lobby organizations in Washington are the U.S. Chamber of Commerce and the Business Roundtable. Also important are the National Association of Manufacturers, which speaks for smaller industrial corporations, and the increasingly influential National Federation of Independent Business, representing small business. The Chamber gains special strength from its ability to mobilize pressure on individual members of Congress through local affiliates. The local capitalist class, which dominates the affiliates, is likely to include important elements of the power structure in a legislator's home district, as well as people who belong to the same social networks as the legislator. When the Washington office wants to pressure senators or representatives on a vote, it can systematically mobilize letters, phone calls, and personal visits from a local business owner, a legislator's former law partner, or a fellow member of the local country club.

The Business Roundtable consists of the CEOs of approximately 200 of the largest corporations. Its power is based on the formidable resources controlled by these corporations and the prestige of those who lead them. The Roundtable typically operates more quietly than the Chamber does. Its stock-in-trade is the personal visit from a CEO and the carefully crafted economic study or legal brief supporting its position. The leaders of the Roundtable have access to members of the

House and Senate and even to the president—entree that no ordinary lobbyist could hope to duplicate. A congressional aide commented, "A visit from a CEO has an unbelievable impact, as perhaps it should. It shows a commitment" (Green 1979:29).

All these business lobby organizations and many smaller business groups collaborated to defeat plans for a national health care insurance system introduced by the Clinton administration in 1993 (Johnson and Broder 1996:194–224, 601–636). Bill Clinton's promise to establish universal coverage helped elect him in 1992. National polls consistently showed that approximately 70% of adults agreed with the proposition that health insurance should be guaranteed to all. Despite this remarkable consensus, the administration's plan never even came to a vote in Congress, and alternative plans to extend coverage were equally unsuccessful.

Organized business spent more than $100 million to fight the administration— enough to finance a national presidential campaign (Johnson and Broder 1996:212). In fact, the opponents hired political professionals experienced in electoral politics and employed the sophisticated tactics of a national election campaign. They mobilized targeted campaign donations, a well-staffed lobbying effort on Capitol Hill, regular public opinion polling, a large-scale television advertising campaign, phone banks, and field operations in key states and Congressional districts. Supporters of national health care (including labor unions) simply did not have the resources to operate on the same scale as the opponents.

The administration's inept political strategy, divisions among Democrats in Congress, and controversial aspects of the Clinton plan itself undeniably contributed to the administration's defeat. But, given popular support for universal health care, a compromised plan might well have emerged from Congress without massive opposition from organized business.

The health care debacle of the early Clinton years can usefully be compared with the major legislative accomplishment of the new Bush administration 8 years later. During his first months in office, Bush pushed through Congress substantial reductions in income and estate taxes over a 10-year period (see Chapter 4). The tax proposals never attracted the broad popular support that national health care had initially enjoyed. Many feared that the loss of tax revenue would undermine important federal programs and understood that the blessings of the tax legislation would mainly flow to the rich. But opponents of the tax bill did not have the resources to mount the kind of campaign that defeated national health care.

The class element in U.S. politics becomes clearest when issues arise, like health care and taxation, that pit a liberal lobby alliance often led by labor unions against a conservative lobby alliance led by business groups. Since the late 1970s, the business alliance has tended to win such confrontations.

Policy-Planning Groups

A step removed from the conflictual world of political campaigns and legislative battles is a quieter and less visible realm of organizations dedicated to formulating and disseminating broad proposals for national policy. Groups such as the Council on Foreign Relations, the Council for Economic Development, and the Business Council are created and financed by the corporate elite, which plays a prominent role in their activities. Corporations have long participated in such organizations through funding and board membership. Domhoff's (1975) unfortunately dated social network analysis of the boards of major corporations, memberships of exclusive social clubs, and participants in policy-planning organizations revealed an elaborate pattern of interconnections.

Similar to the policy groups, both in their functions and their links to the national upper class, are the major charitable foundations and the policy research "think tanks." Foundations such as Rockefeller, Ford, Lilly, and Kellogg (all named for the wealthy families that endowed them) fund research and pilot projects to test policy ideas. Many of the best-known think tanks are clustered in Washington, where they can feed their research findings and policy recommendations to sympathetic politicians, lobbyists, and journalists. Among the best-known think tanks are the Brookings Institution, which has particularly influenced Democratic policy makers, and the Heritage Foundation and the American Enterprise Institute (AEI), which are influential among Republicans. One key function of the policy groups, foundations, and think tanks is to back the careers of public policy intellectuals, many of whom are channeled into government positions.

Indirect Mechanisms of Capitalist-Class Influence

Capitalist-class influence over government is not limited to the direct means we have been describing (recruitment to decision-making positions, campaign financing, lobbying, and domination of policy-planning institutions). The capitalist class can also affect government policy indirectly, through its control of the economy and the mass media. A defining characteristic of a capitalist society is the existence of a relatively small class that controls most productive wealth and therefore independently makes investment decisions that can decisively affect the welfare of other classes. Although governments in capitalist societies have limited control over what business leaders do, their political fortunes are closely linked to business decisions. The connection is often described in terms of a "business confidence." If business leaders lack confidence in a government or its policies, they are not likely to risk their capital in new investments. The resulting decline in aggregate level of investment will soon be reflected in a rising level of unemployment, which, in turn, will subject the government to pressure from an electorate dissatisfied with the state of the economy.

If the government wants to rectify the situation without making basic changes in the capitalist economic order, it must find a way to regain the confidence of investors. What sorts of government policies are likely to alienate business confidence? Basically, any that threaten business profits, from "excessive" taxation of corporate income to the imposition of expensive regulations designed to reduce pollution or guarantee worker safety. The precise factors that lead to a loss of confidence are less important than the essential fact that this mechanism gives the capitalist class an indirect veto over government policy.

Two implications of the business-confidence veto are particularly worth noting. One is that it can influence a government without actually curtailing investment. The mere risk of such action is enough to persuade decision makers to reconsider a proposed policy or the appointment of a cabinet officer whose opinions might sound threatening to business. The possible effect of government action on investor behavior is frequently raised as an issue in public policy debates. The other implication is that the veto mechanism does not require conscious, concerted action by members of the capitalist class to be effective. Individual investment decisions, based on objective assessment of potential risk and profitability, can collectively produce a downturn in business activity and subject a government to popular pressure (Bloch 1977).

Governments are also subject to the limits imposed by private control of the mass media, through which people receive information about public affairs. In the United States, virtually all significant media are owned by the local and national capitalist classes (though they might just as well be organized as cooperatives, like the respected French paper *Le Monde;* as semiautonomous public bodies, like the British Broadcasting Company; or as organs of political parties, like a number of European papers). Control of the media has become highly concentrated, and the principal media organizations are themselves typically major corporations or are owned by major corporations. The most notable exception to this rule is the Public Broadcasting Service (PBS) and National Public Radio, which only have a small portion of the resources or audience of the commercial networks.

In a world of multiplying information channels, most Americans who pay any attention to national and international affairs are still likely to receive their news from one of five TV networks: ABC, CBS, NBC, CNN, and Fox. All are owned by large media conglomerates, except NBC, which is owned by General Electric. The companies that publish the two most influential daily newspapers, the *New York Times* and the *Washington Post,* and the two major news magazines, *Time* and *Newsweek,* are also among the country's largest corporations. A dozen or so newspaper chains account for more than half of the country's daily newspaper circulation. Most American newspapers get their national and international news from a single source, the Associated Press (AP), a cooperative owned by the media it serves (Dye 1995:113).

Capitalist-class influence over the media is not limited to the power of ownership. Because the media are operated for private profit and most of their income comes from corporate advertising, media managers are sensitive to pressure from advertisers. The networks are also subject to the influence of the affiliated stations that broadcast their programs to local audiences. Early in the history of television broadcasting, Edward R. Murrow's brilliant and controversial current affairs program *See It Now,* carried by CBS, was forced off the air after its corporate sponsor withdrew and no regular replacement could be found. In the 1950s, programs that dealt with the issue of racial discrimination or employed black actors could not appear on network television because corporate sponsors refused to be associated with them, and some affiliates (particularly in the South) refused to carry them. Today, there are few confrontations between advertisers and networks. These early confrontations established unwritten standards that continue to guide commercial television broadcasting. As an executive of a major advertising agency explained, direct interference by the advertiser is rare "because the producers involved and the writers involved are normally pretty well aware of what might not be acceptable" (Barnouw 1978:54; see also Tuchman 1974).

Of course, to sell advertising, commercial media must attract an audience, and they are reluctant to offend or bore that audience with unattractive content. But as the ad man's comment suggests, the main power that the capitalist class exercises over the media is the power to impose implicit limits on what is "acceptable." The media, in turn, operate on their audiences not by imposing specific ideas but by defining the subjects that are appropriate for consideration and delineating the range of reasonable opinion. In other words, they help define the public agenda. Their ability to do so is increased by the concentration of media control. Thus, until the late 1950s, racial inequity was not a national issue, though it was most certainly a serious national problem. For decades the issue of national health care was invisible, though it certainly was a serious problem for many Americans.

The Capitalist-Class Resurgence

Who has the power? This is the question that pluralists, elitists, and class theorists are trying to answer in the debate we examined at the beginning of this chapter. The question assumes that the distribution of power is stable—it does not vary over time. But this is probably not a safe assumption. It seems clear, for example, that the capitalist-class and business interests in the United States have gained at the expense of other competitors for power during what we have called the Age of Growing Inequality.[17]

In the early 1970s, there was a growing sense of vulnerability and declining power in capitalist circles. The economic system was changing in ways that seemed threatening and unpredictable. Wages had been rising, profits had been stagnating, and productivity had been declining. The international economy, dominated by the United States since the end of World War II, was becoming much more competitive. The U.S. economy was twice shaken in the 1970s by abrupt leaps in the world price of oil. Increasing government regulation in areas from environmental practices to consumer protection and workplace safety seemed to be raising the cost of doing business. Many business leaders felt politically isolated. John Harper, then chairman of Alcoa, recalled the period from the happier perspective of the 1980s:

> We [corporate leaders] were not effective. We were not involved. What we were doing wasn't working. All the polls showed business was in disfavor.
>
> We didn't think that people understood how the economic system works. We were getting short shrift from Congress. I thought we were powerless in spite of the stories of how we could manipulate everything (quoted in Blumenthal 1986:77).

---

[17]See Blumenthal 1986; Blumenthal and Edsall 1988; Edsall 1984; Ginsberg and Shefter 1990.

Since the early 1970s, business leaders have taken a more direct and aggressive role in national politics. Corporations, as we have seen, seized the opportunity presented by the PAC provisions of the post-Watergate reforms. Business lobbyists perfected the upscale "grassroots" campaign—mobilizing local business leaders, stockholders, depositors, suppliers, or dealer networks to influence Congress. In 1972, Alcoa's Harper joined other top corporate leaders to found the Business Roundtable. About the same time, the Heritage Foundation was started, with the help of a $250,000 donation from Colorado brewer Joseph Coors, and the American Enterprise Institute began its transformation from an inconsequential research center into a key player in public policy. Both received financial backing from major corporations and foundations endowed by wealthy families such as the Mellons, the Pews, and the Olins (Blumenthal 1986; Edsall 1984:Chapter 3).

These efforts to reassert the power of the privileged have paid off. In the late 1970s, business won a series of key legislative battles—for example, defeating both the consumer protection agency bill and labor reform legislation that would have made it easier for unions to organize workers. In Chapter 4, we saw that federal taxes on high incomes, inherited wealth, and corporate earnings have been sharply reduced since the early 1970s. The real incomes of workers have declined, and the distribution of income has become more concentrated. In the next chapter, we will see that the power of organized labor declined as the power of business grew.

The power shift that began in the 1970s contributed to Republican victories in five out of seven presidential elections from 1980 to 2004, and the election of conservative Republican majorities to both houses of Congress from 1995 to 2006, for the first time in decades. However, there is no reason to assume that the new power structure is unalterable. The changes that have taken place might themselves set off further change, in directions we cannot now imagine.

Conclusion

We began this chapter by defining elite, pluralist, and Marxist perspectives on power. We then took a close look at C. Wright Mills's *Power Elite*, his pluralist and Marxist critics, and some recent elitist conceptions of the structure of power in America. Mills emphasized the growing scale of corporate, government, and military organization and the corresponding concentration of national power in the "corporate rich," top government officials, and the leaders of the military. Much of the criticism of Mills centered on the question of "elite cohesion." Do the members of a putative elite act together in pursuit of common objectives and in opposition to other groups? Mills did point to mechanisms that tended to unify the members of his power elite, including similar social backgrounds, shared elite education, association through upper-class society, movement of personnel between elites, and the common experience of managing large organizations. But the pluralists were unconvinced. Where Mills saw one cohesive elite, the pluralists saw many competing veto groups.

On the other hand, Marxist critics of *The Power Elite* thought Mills had been altogether too successful in demonstrating cohesion, but the key unifying force was the dominant corporate sector within the elite. America was not ruled by a power elite but by a corporate-based capitalist class. The contemporary writers on elites we reviewed agreed with Mills on one thing: power is concentrated in large organizations and the elites that control them. Beyond that point, their accounts of the system diverged, especially on the issues surrounding cohesion.

Our extended examination of the national capitalist class established the following: A small class controls most corporate stock, and although major corporations are typically run by their top executives, the interests of corporate stockholders and top managers are aligned. We therefore consider these executives part of the capitalist class. There is apparently—our information on this is somewhat dated—a significant overlap between the national capitalist class and the national upper class represented by such institutions as the *Social Register* and exclusive social clubs and prep schools; the upper class provides at least some of the social glue that binds the members of the capitalist class together. Finally, the national capitalist class has powerful means to shape national politics; these include placement of its members in top decision-making positions, campaign financing, lobbying, creation of policy-planning organizations, exercise of the business-confidence veto, and control of the public agenda through the mass media. Looking back over the lasts three decades, we concluded that the power of the capitalist class has grown, relative to other classes.

A pluralist would be quick to point out that neither the formidable array of political resources available to the capitalist class nor recent indications of growing capitalist-class power are definitive proof of domination by that class. We make no such claim. But we are committed to examining how social classes participate in the political system and how the balance of power between them has shifted in the Age of Growing Inequality. With those goals in mind, we amplify the picture we have painted here in the next chapter, which deals with class consciousness and conflict between classes in electoral and industrial contexts.

Key Terms Defined in the Glossary

capitalist class
chief executive officer (CEO)
class perspective (see elite;
   pluralistic perspective)
elite
elite cohesion
elite perspective
   (see pluralist perspective)
endogamy

Establishment, the
pluralist perspective
political action committee
   (PAC)
power
privileged classes
soft money

Suggested Readings

Bottomore, Tom. 1966. *Elites in Modern Society.* New York: Pantheon.
> *Short, lucid survey of elite theory.*

Cookson, Peter and Caroline Hodges Persell. 1985. *Preparing for Power: America's Elite Boarding Schools.* New York: Basic Books.
> *How the upper class is educated.*

Domhoff, G. William and Hoyt B. Ballard, eds. 1968. *C. Wright Mills and the Power Elite.* Boston: Beacon.
> *Excellent set of critical essays on Mills's power elite thesis.*

Dye, Thomas R. 2002. *Who's Running America? The Bush Restoration.* 7th ed. Englewood Cliffs, NJ: Prentice Hall.
> *The national elite, sector by sector.*

Frank, Robert. 2007. *Richistan: A Journey Through the American Wealth Boom and the Lives of the New Rich.* New York: Crown.
> *A well-informed examination of the fortunes, lives, and politics of the new rich, by a reporter who has covered them for the* Wall Street Journal.

Graetz, Michael J. and Ian Shapiro. 2005. *Death by a Thousand Cuts: The Fight Over Taxing Inherited Wealth.* Princeton, NJ: Princeton University Press.
> *How a small, well-financed group turned the estate tax into the death tax and promoted legislation to abolish it.*

Graham, Lawrence Otis. 1999. *Our Kind of People: Inside America's Black Upper Class.* New York: HarperCollins.
> *A sophisticated insider's account of the black upper class.*

Johnson, Haynes and David Broder. 1996. *The System: The American Way of Politics at the Breaking Point.* Boston: Little, Brown.
> *An engaging, thoughtful account of the Clinton administration's failed attempt to create a national system of universal health care.*

Judis, John B. 1991. "Twilight of the Gods." *Wilson Quarterly* 5 (Autumn):43–57.
> *Intriguing account of the rise and fall of the "American establishment" of bankers, corporate lawyers, and scholars who once made U.S. foreign policy. Helpful annotated bibliography.*

Malbin, Michael J., ed. 2006. *The Election After Reform: Money, Politics, and the Bipartisan Campaign Reform Act.* Lanham, MD: Rowman & Littlefield.
> *A useful guide to recent changes in campaign finance laws and their effects.*

Ostrander, Susan. 1984. *Women of the Upper Class.* Philadelphia: Temple University Press.
> *The lives of upper-class women and their roles in the maintenance of their class.*

Phillips, Kevin. 2002. *Wealth and Democracy: A Political History of the American Rich.* New York: Broadway Books.
> *Concentrated wealth, democracy, and the tensions between them in American history since colonial times. Compares current era of concentrated wealth and power with previous gilded ages.*

CHAPTER 9

Class Consciousness and Class Conflict

I believe that leaders of the business community, with few exceptions, have chosen to wage a one-sided class war today in this country.

Douglas Fraser, president of the United Auto Workers (1978)

W e owe the concept of class consciousness to Karl Marx. Its role within his theory was pivotal, joining individual experience to broad social structures and transforming alienated individual resentment of the capitalist present into decisive striving for the socialist future.

Class consciousness implies an awareness of membership in a group defined by an economic position, a sense that this shared identity creates common interests and a common fate, and, finally, a disposition to take collective action in pursuit of pursuit of class interests. At some points in his work, Marx implied that only a group whose members experience such a consciousness can be defined as a class. Elsewhere, he carefully distinguished between a *class-in-itself* and a *class-for-itself.* The first is a class in a formal, definitional sense: Its members share an objective class position (defined by the analyst) but are unaware of their common situation. The second is a class in an active, historical sense: Its members are aware of common interests; they engage in militant action focused on goals that they conceive as being in direct opposition to those of other classes. Thus, embodied in Marx's conception of class consciousness—especially in the notion of a class-for-itself—is the expectation of class conflict.

In its fullest sense, class consciousness is not just an aspect of public opinion ("What percentage of blue-collar workers supported the candidate?"), but an intense, collective involvement in the events of a critical historical juncture. It develops out of a long series of strikes against bosses who exploit workers and riots against authority that brutalizes the masses. It culminates in urban mobs roaming the streets and burning the buildings that symbolize upper-class domination and in peasants seizing the land they work, and it ends with a revolutionary seizure of power in the name of the oppressed: Paris in 1871, Mexico in 1910, Moscow in 1917, Peking in 1949, Havana in 1959.

Revolution is rare, but simmering class struggle is common. Slave revolts, violent strikes, local mobs on a rampage—these occur regularly in many societies. More institutionalized and controlled forms of class struggle, such as union organizing campaigns and political movements that seek legislative power to help the underprivileged, are considered a normal and healthy part of a democratic society. Historians study past revolutions; sociologists usually focus on the initial development of class consciousness, which could, under very specific historical circumstances, lead toward revolutionary consciousness but is more likely to result in peaceful change or historical stagnation. In this chapter, we examine the extent to which people are aware of sharing a class identity and class interests, the social factors that advance or retard the development of this consciousness, and its relationship to political opinion and behavior. We will also focus on class conflict as reflected in two arenas: electoral politics and labor relations.

Much of the material in this chapter examines this basic causal sequence:

$$\text{Objective Class Position} \longrightarrow \text{Class Consciousness} \longrightarrow \text{Class-Oriented Political Behavior}$$

Here, class position is objective in the sense that it is determined from outside by the analyst in contrast to the individual's subjective consciousness of his or her class. In this sequence, class consciousness links objective class position (or, more generally, the individual's economic situation) and political behavior. Thus, we expect most factory workers (objective position) to think of themselves as members of the working class (class consciousness) and to behave accordingly when they vote, strike, join the revolution, or just decide how they feel about a specific political issue (political behavior).

We need not assume that class position can only influence political behavior through class consciousness. We may find (in the language we applied to social mobility) other causal "paths" linking the two. For example, a worker who doesn't especially think of himself as working class might nonetheless vote for a working-class party because his relatives and the other people he knows in his working-class neighborhood are doing so. We will explore both of the causal arrows in the diagram on the previous page, with questions like these: How likely is class situation to be translated into class consciousness? Under what conditions does this happen? How significant is the influence of class consciousness on political behavior?

Marx and the Origins of Class Consciousness

One of Marx's major objectives was to isolate the social conditions that encourage class consciousness. He hoped that, by understanding the process, he could determine how to intervene and accelerate it. Specifically, he asked, What inherent tendencies of capitalist society are likely to produce a class-conscious proletariat? Here are the factors that he regarded as especially significant:

1. *Concentration and communication.* The process of industrialization in capitalist society concentrates the proletariat in big cities, working-class neighborhoods, and large factories. This process promotes communication among workers, leading to a recognition of common problems and facilitating efforts at political organization.

2. *Deprivation.* Marx expected a progressive impoverishment of the proletariat, if not in an absolute sense, at least relative to the rising productive capacity of the industrial economy and the wealth of the bourgeoisie.

3. *Economic insecurity.* Marx was convinced that the proletariat's sense of deprivation would be exacerbated by the periodic experience of unemployment during the downturns in the capitalist economy, which, he observed, is quite subject to boom-and-bust cycles.

4. *Alienation at work.* Marx identified the mindless, repetitive, unsatisfying quality of factory-type labor with capitalism. (For a modern rendition of this argument, see Braverman 1974.) Such labor is fundamentally at variance with human nature as Marx understood it and is therefore a spur to the development of class consciousness.

5. *Polarization.* The swings of the capitalist economy drive smaller enterprises out of business; their owners are forced into the proletariat, and control of the

economy becomes further concentrated at the top. The result is the steady deple-
tion of the middle ranks and the corresponding development of a society polarized
between a tiny, wealthy bourgeois minority and an impoverished proletarian
majority.

6. *Homogenization.* Within the proletariat, Marx observed a lowering of skill
levels and therefore an equalization of wage levels produced by adaptation to the
simple requirements of machine tending in the modern factory. This tendency
leads to a less stratified, more homogeneous proletariat, which, because of its
shared condition, is more disposed to unified political action.

7. *Organization and struggle.* To defend itself, the proletariat is drawn increasingly into working-class parties and labor organizations. Marx believed that participation in such organizations and the experience of struggle against capitalist
employers (backed by the bourgeois state, with its police and armies), would promote the development of a revolutionary class consciousness.

These factors are, by and large, related to the first element in the causal sequence
diagrammed above: objective class position. They refer to the way Marx believed
capitalist development was shaping the class system (polarization, homogenization,
concentration) or to workers' conditions of employment (insecurity, deprivation,
alienated work). The last factor (organization and struggle) is an exception. It suggests that political behavior can react back on class consciousness.

The revolutions that Marx's theory anticipated in the advanced industrial
countries never came. However, class-based revolutions in industrializing agrarian
states (Mexico, Russia, and China, for example) were a characteristic feature of
twentieth-century history. In the industrial nations, working-class parties and labor
movements reshaped political systems and economic life. In both cases, the factors
isolated by Marx have proven important. For example, international studies of
working-class support for leftist parties have found that workers who live in big
cities, work in big plants, hold low-skill jobs, or experience unemployment are
especially likely to vote for the left (Lipset 1960; Szymanski 1978). The findings are
consistent with Marx's emphasis on three of the factors listed earlier: concentration,
alienating work, and economic insecurity.

Understood as variables operating under specific circumstances, the seven
Marxian factors can even help us understand the failure of revolution in the
advanced countries. For example, Marx expected parallel processes of class polarization and homogenization in capitalist society. Property relations did become
polarized: productive property was increasingly concentrated in the hands of a
small minority. Nevertheless, homogenization of the proletariat was undercut by
occupational differentiation and the corresponding spread in incomes, even among
manual workers (see Chapter 3). This process limited the sense of common identity and the shared experience that are the bases of class consciousness. We might
say that Marx was correct in identifying the key sociological processes, even if he
was historically wrong in predicting their outcome. Much of this chapter is devoted
to applying his best insights to social and political circumstances he could not have
imagined.

Richard Centers and Class Identification

Contemporary interest in the concept of class consciousness is based on the idea that it connects objective class position (measured by occupation, income, or wealth) and political behavior. That is, we (like Marx) assume that people who recognize and articulate their class position are more likely to promote their class interests. A systematic and sustained effort to investigate this linkage in American society grew out of the work of Richard Centers (1949), who focused on one aspect of class consciousness: class identity, the sense of belonging to a particular social class.

Centers began by noting that in previous public opinion surveys (such as the famous one conducted by *Fortune* magazine in 1940), about 80% of Americans called themselves middle class. Some popular writers seized these figures to proclaim that America was almost completely a middle-class country—that if Americans had any class consciousness at all, it simply meant that they mostly thought of themselves as belonging to the same big group. But Centers noticed that the figure quoted came from the following survey item, which offered only three alternatives:

What social class do you consider that you belong to?
1. Upper class
2. Middle class
3. Lower class

He also noticed that when respondents were asked the question in open-ended form (without a specific list of answers from which to choose), many called themselves working class. Centers made a reputation for himself by adding working class to the reply alternatives offered by the *Fortune* survey. When he asked a nationally representative cross section of adult white men which of the four classes they belonged to, the responses were radically different from those to the 3-choice question posed by *Fortune*. Now the majority of respondents chose the label working class (see Table 9.1).

Table 9.1   Class Identification

	In Percent	
	Centers (1949)	GSS (2000)
Upper Class	3	3.8
Middle Class	43	45.2
Working Class	51	45.4
Lower Class	1	5.0
Don't know/Other	2	0.5
Total	100	100
N	1097	2817

SOURCES: Centers 1949 and author's tabulation from General Social Survey 2000 data.

Centers rightly concluded that Americans do not like the term *lower class* and that this attitude was the main conclusion to be drawn from the *Fortune* survey, not that most Americans thought of themselves as middle class. His confidence in the results of the surveys was strengthened because only a tiny minority of respondents refused to accept one of the labels suggested in the question. Even fewer were inclined to deny the existence of social classes. Centers (1949) was moved to say, "The authenticity of these class identifications seems unquestionable" (p. 78).

Variants on Centers's class identification question have been used in numerous surveys over the years and gotten roughly similar results. Table 9.1 shows the responses to the Centers item in the 2000 General Social Survey, a periodic national survey of adults. The most notable difference in 2000 was the higher, though still small, percentage of adults willing to accept the label "lower class"—reflecting, perhaps, the growing number of Americans who feel themselves falling behind the mainstream.

Despite the remarkable durability of his class identification item, Centers's claim of "unquestionable authenticity" for these responses is not quite justified. Surveys that ask for class identification without suggesting answer alternatives with an open-ended question produce less orderly results. When they are not given fixed alternatives, people invent dozens of class labels for themselves beyond Centers's basic four. We cannot, then, use the answers to Centers's "forced choice" question as literal descriptions of the way people freely conceive of their own class positions. We can, however, assume that the forced-choice question is in some sense a measure of class consciousness because, as we will see, responses to it are related to both class position and political attitudes in the sense suggested by the causal diagram presented earlier.

Correlates of Class Identification

What types of persons chose the particular labels offered in class identification surveys? Centers regarded occupation as the principal basis of class identification. When he sorted respondents by occupation, he found that 70% or more of professionals and businessmen considered themselves middle class, while more than 70% of manual workers considered themselves working class. Later studies produced similar results. Centers's data conform to a pattern that is by now familiar: The results were fairly clear-cut at the extremes of the class structure but somewhat ambiguous in the middle. In particular, more than a third of sales and office workers among Centers's respondents and about half of that group in subsequent studies labeled themselves working class (Centers 1949:86; Hamilton 1975; Schreiber and Nygreen 1970).

Centers's view that occupation is the main determinant of identification is supported by an analysis of national survey data by Hodge and Treiman (1968). They found that occupation (of the family's main earner) was a stronger predictor of class identification than either family income or respondent's education, but the three variables considered together left much of the variance in class identification unaccounted for. In other words, objective class position seemed to be a crucial but not a decisive determinant of this aspect of class consciousness. Hodge and Treiman found one additional factor that was significantly and independently related to class identification: *association*—one of our basic class variables. The class positions of friends, neighbors, and kin were strong influences on the formation of class identification.

People with largely high-status associations, whatever their own position, were more likely to identify as middle class. Likewise, those with predominantly low-status associations were more likely to think of themselves as working class. Put differently, your class identification depends partly on your objective class position and partly on whom you know.

Married Women and Class Identification

Centers's original class identification surveys referred only to men, but later surveys included women. If we are interested in the relationship between objective class position and identification, the inclusion of women raises an intriguing question: Does a working wife base her class identification on her husband's occupation, her own occupation, or some combination of the two?

Recent research suggests that the class identification of working wives is influenced by both their own and their husband's occupations—though the influence of the husband's job is typically stronger (Sorensen 1994:34–35). Beeghley and Cochran (1988) wondered why this might be true. Their analysis of data from several national surveys concludes that the key factor is the wife's attitude toward gender roles. Working women who gave traditional answers to survey questions about gender roles (for example, they said a wife should not work if her husband could support her) based their class identity on their husband's occupation. Those with more egalitarian views considered their own and their husband's jobs, along with other class factors pertaining to both spouses. However, the effect of wives' attitudes on the overall pattern of class identification for men and women is probably modest. The reason is that wives and husbands tend to have broadly similar jobs. Not too many lawyers and CPAs are married to construction workers.

We can speculate that spouses answering the class identification question are not simply weighing two occupations but are thinking of a standard and style of living, a set of associations, and particular values and attitudes that are shared by the members of a household. Traditionally, all these things depended largely on the husband's job, and it was reasonable to assume that the class positions of most individuals were socially fixed by the characteristics of the (male) heads of their families. But rising female participation in the labor force, growing family dependence on dual incomes, and corresponding changes in gender norms undermine this conventional wisdom. Research on the class identification of working wives is consistent with the idea that family continues to be the relevant basis of class position, but the standing of families is no longer solely dependent on the activities of men.

Class Identification, Political Opinion, and Voting

Our interest in subjective class identification, like our concern with class consciousness generally, stems from the idea that consciousness, as we suggested at the beginning of this chapter, is a critical link between objective class position and political attitudes or behavior. If there is anything to the notion of class identification, we would, for example, expect low-status workers who identify themselves as working class to take fairly liberal positions on economic issues and those who identify themselves as middle class to take somewhat more conservative positions. We would expect a similar pattern in candidate preferences in elections. Centers was able to show, using his own data, that political opinions were, in fact, affected by the class identification.

We tested the influence of class identification on political ideology with an item from the 2000 General Social Survey. The survey asked whether respondents agreed or disagreed with the idea that it was "all right" for some to accumulate great wealth while many others live in poverty. About half of the respondents who expressed an opinion disagreed—that is, they thought such differences of wealth and poverty were not "all right" (see Table 9.2). As the figures in the "All" column indicate, objective class position (here measured by income) strongly affected reactions to this issue: People in the bottom class were 23% more likely to disagree with the statement than were those in the top class. But class identification was also important. For example, lower-income respondents who identified as "working class" were 10% more likely to disagree than were lower-income respondents who identified as "middle class." Opinion here is influenced both by (objective) income class and (subjective) class identification.

Table 9.2   Income, Class Identification, and Opinion

"In a free society it is all right if a few people accumulate a lot of wealth and property while many others live in poverty." *Percent disagree.*

| | | Class Identification | |
	All	*Working*	*Middle*
Income Class			
Under $25,000	63	66	57
$25,000–$50,000	55	54	55
Over $50,000	40	48	37
Total	53	58	47
N	967	500	457

SOURCE: Author's tabulation from 2000 General Social Survey.

NOTE: Total disagree and strongly disagree among those expressing opinion. Working class includes lower class; middle class includes upper-class identification (see Table 9.1).

Bott: Frames of Reference

Class identification gives us an indirect way of measuring class consciousness and estimating its political importance. But class identification tells us little about how people actually turn their experience of the class structure into some notion of their own place within it. Elizabeth Bott (1954) has pointed out that "people do not experience their objective class position as a single clearly defined status" (p. 262). We might add that such clear definitions are the result of calculated decisions on the part of academic researchers; they create concepts such as "upper-middle class" or "bourgeoisie" and through hard thinking attach some specific empirical criteria for membership. (Granted, they may start with words in popular usage, but by the time they have finished their ratiocinations, the original words have taken on new meanings.) Naturally, they endow their concepts with connotations that derive from the researchers' own general philosophy; thus, Warner thought of prestige strata, and the Marxists thought of actual or latent conflict groups. Then the researchers go into the field and try to discover the degree to which the populace thinks as the concepts suggest they should, and the investigators feel a growing sense of triumph the more closely they can fit the data to the concepts.

But, said Bott (1954),

> When an individual talks about class, he is trying to say something, in a symbolic form, about his experiences of power and prestige in his actual membership groups both past and present. These membership groups—place of work, friends, neighbors, family, etc.—have little intrinsic connection with one another, especially in a large city, and each of the groups has its own . . . system of prestige and power. When he is comparing himself with other people or placing himself in the widest social context, he manufactures a notion of his general social position out of these segregated group memberships. . . . The group memberships are not differentiated and related to one another; they are telescoped and condensed into one general notion. (P. 262)

The average man or woman is aided in conceptualizing by ideas and terms that have diffused into popular culture from intellectual debate. Thus, especially in Europe, many factory workers have long been subjected to propaganda that stems from Marxist ideology. Naturally, they not only use class-conflict terminology, they also perceive their own position and interpret their everyday experiences in terms of conflict. Similarly, the American middle classes have been fed a very different line of propaganda, emphasizing individualism and the American dream of social mobility. Consequently, they tend to view in individual terms experiences that a European might interpret from a class perspective.

Therefore, any individual's self-perception in a stratification order is a combination of (1) actual experiences in a wide variety of contexts in many membership groups and (2) verbal theories about society, which are usually vague and somewhat contradictory commonsense notions that have filtered down from the theorizing of intellectuals and propagandists. Bott (1954) emphasizes that the individual engaged in this process of synthesis is "an active agent [who] does not simply internalize the norms of class which have an independent external existence (p. 263).

Consequently, the social reality of identification that we are studying is complex rather than simple, and when we simplify it (as we must for certain purposes) into categories such as middle class or working class, we are straying from the original facts. The simpler and neater the scheme, the further it is from reality.

Elections and the Democratic Class Struggle

Although no advanced industrial country has experienced the convulsive class revolution envisioned in the *Communist Manifesto,* most have passed through periods of bitter class confrontation and continue to experience less dramatic, institutionalized struggles over conflicting class interests. Class conflict is especially evident in two realms: electoral politics and labor relations. Most of the remainder of this chapter will be devoted to these areas.

Elections in modern democracies have been characterized as manifestations of "democratic class struggle" in recognition of the representative role of political parties (Anderson and Davidson 1943; Lipset 1960:Chapter 7). Synthesizing the available evidence in 1960, Seymour Martin Lipset wrote,

> Even though many parties renounce the principle of class conflict or loyalty, an analysis of their appeals and their support suggests that they do represent the interests of different classes. On a world scale, the principal generalization which can be made is that parties are primarily based on either lower classes or the middle and upper classes. (P. 230)

In most parliamentary systems, parties can be arrayed on a spectrum from right to left, with the former upholding the interests of the privileged classes, and the latter attacking them on behalf of the less fortunate. In Great Britain, for example, the Labor Party has traditionally drawn working-class support, while the Conservatives have run strong among managers and professionals. In France, manual workers lean toward the Socialists (and, in the recent past, the Communists); business owners, executives, and professionals and farmers tend toward the right-wing parties.

The relationship between classes and parties has traditionally been looser in the United States than in most Western democracies. Democratic political systems typically have at least one major party that identifies itself as socialist and presents itself as a partisan of the working class (for example, Socialist, Social Democratic, and Labor parties of Western Europe). The Democratic Party in the United States has traditionally been regarded as the party of the "common man," but it has never called itself socialist, and it has become increasingly coy about appealing directly to working-class interests. Nevertheless, the Democrats have done better among working-class voters than among middle-class voters for as long as anyone has bothered to keep track.

Exit polling in two recent elections shows a strong relationship between income and party vote. In the 2004 presidential election that pitted Republican George W. Bush against Democrat John Kerry and the 2006 elections for the House of

Representatives, the Democratic percentage of the vote was higher at successively lower income levels (Table 9.3). In 2006, for example, there was a nearly 20% gap in support for Democrats between the lowest fifth of voters (those below $30,000, according to the exit poll) and the top fifth (above $100,000 in the poll).

Despite these differences by income, the percentage gap between upper–white-collar and blue-collar voters in support for the Democrats has typically been modest and smaller than corresponding differences in European democracies.[1] Explanations for this so-called "peculiarism" of the American party system often revolve around the role of ethnicity, religion, and geography in our politics. Blacks, Jews, and, to a lesser degree, Catholics are more likely to vote Democratic, whatever their social class, than are white Protestants. Regional preferences have shifted over time, especially in the South, but have tended to dilute the class pattern in elections.

The Democratic Party that emerged from the New Deal under Franklin Roosevelt in the 1930s was a party with strong working-class support but was not a working-class party. It could mobilize working-class voters with the help of the labor movement, which had grown rapidly under the New Deal, and the party's big-city political machines. But ethnicity blurred class lines. Most white Protestant workers supported the Democrats, but some of them (more than among Jewish or Catholic workers) favored the Republicans. The Democratic Party also had important upper-class backers, drawn from the "ethnic rich"—men such as John F. Kennedy's multimillionaire father, Joseph, a Catholic, or Jewish financier Bernard Baruch. The ethnic rich bankrolled the party, and, although they were probably more ideologically flexible than their Protestant counterparts in the Republican Party, they were nonetheless a conservatizing influence on their own party.

Table 9.3   Party Preference in 2004 and 2006 Elections by Income

| | *In Percent* | | | | | |
| | *2006 U. S. House* | | | *2004 Presidential* | | |
Household income	**Dem.**	**Rep.**	**Total**	**Dem.**	**Rep.**	**Total**
Under $15,000	69	31	100	64	36	100
$15,000–$30,000	63	37	100	58	42	100
$30,000–$50,000	57	43	100	51	49	100
$50,000–$75,000	51	49	100	43	57	100
$75,000–$100,000	53	47	100	45	55	100
$100,000–$150,000	48	52	100	42	58	100
$150,000–$200,000	48	52	100	42	58	100
$200,000 & Above	46	54	100	36	64	100

SOURCE: CNN.com.

NOTE: Percentages are of voters who expressed preference for major party candidates. Excludes those who did not state preference in exit poll or indicated minor candidate. Family incomes in current dollars, not inflation adjusted.

---

[1]For some comparisons, see Charlot 1985; Daalder and Koole 1988; and Frears 1988.

The New Deal coalition of working-class and ethnic voters enabled Democrats to dominate American politics for decades. But beginning with Richard Nixon's 1968 victory, the Republicans aggressively challenged the political status quo, often winning the presidency and control of one or both houses of Congress. At the same time, the New Deal coalition was losing internal cohesion. The result, apparent by the 1980s, was not a decisive realignment in national politics such as occurred under Roosevelt but, rather, an inconclusive standoff, epitomized by the 2000 elections, which produced a virtual tie in the presidential contest and a 50/50 split in the U.S. Senate.

This remaking of national politics was connected with important changes in American society. In the prosperous years after World War II, many of the children and grandchildren of immigrants moved into the middle and upper–middle classes. Even if they retained their Democratic affiliation, these people often became more conservative and more open to the political message of the Republicans. They were increasingly likely, for example, to see themselves as beleaguered taxpayers rather than as beneficiaries of government programs. The class differentiation of the descendants of immigrants convinced Democratic leaders to dilute their party's already weak working-class identity. As second- and third-generation Americans moved to the suburbs, the urban Democratic political machines they had supported went into decline, undercutting the party's ability to turn out voters.

Over time, Democratic positions on issues such as Vietnam, affirmative action, welfare, abortion, and gun control have strained the loyalty of working-class supporters. These issues, which were ably exploited by the Republicans, divided a large sector of blue-collar Democrats from the party's upper–middle class activists. Affirmative action issues in the workplace split black and white working-class Democrats, especially under the difficult economic conditions that prevailed in the 1970s and 1980s. The Democrats lost much of their traditional white support in the South because of the party's identification with the cause of civil rights. More recently, a strong, conservative, fundamentalist Christian movement focused on social issues including abortion and homosexuality has wrested working-class support from the Democrats.

Finally, the relative weights of business and labor in national politics began to change in the 1970s. In response to an increasingly competitive economic environment, organized business interests began to lobby more aggressively, give more money to politicians, and fight the labor movement more directly. In a shifting economy, the proportion of unionized workers in the labor force was declining—a trend accelerated by the policies of Republican administrations. Organized labor had been one of the mainstays of the New Deal coalition. Now its ability to mobilize blue-collar voters for the Democrats, its capacity to finance political campaigns, and the strength of its voice in Washington were all waning. Under these circumstances, Democratic officeholders inevitably became less attentive to the needs of organized labor.

In sum, the class differentiation of the descendants of immigrants, the emergence of issues that divided traditional Democratic constituencies, the decline of labor, and the gains of organized business interests in national politics all contributed to the decline of the New Deal coalition and further weakened the representation of working-class interests in American politics.

Social Class and Party Identification and Support for Social Programs

Despite the weakening of working-class representation in American politics, working-class Americans still tend to identify with the Democratic Party, as Table 9.4 demonstrates. Except for those in the top class, which appears evenly split, Americans are inclined to favor the Democrats. Their strongest support comes from the bottom (lower manual) class. Support for social programs and for liberal positions on economic issues like the minimum wage is also strongest at lower class levels.

The 2000 National Election Survey (NES) provides some recent evidence. The survey included items asking whether spending on a series of federal programs should be "increased, decreased, or kept about the same." The results (Table 9.5) indicate that majorities of Americans favor increased spending on schools, child care, social security, and "aid to the poor." Support was strongest at the bottom and weakest at the top of the occupational class structure. Respondents were less enthusiastic about "welfare programs" than "aid to poor people," a demonstration of how easily language can skew survey results (T. Smith 1987). Either way, a class gradient was evident in responses.

Table 9.4   Occupational Class and Party Identification

	Occupational Class Percent			
	Upper White Collar	Lower White Collar	Upper Manual	Lower Manual
Party				
Democrat	45.5	50.7	49.4	66.0
Republican	46.4	37.2	39.1	19.9
Independent	8.1	12.1	11.5	14.1
Total	100	100	100	100

SOURCE: Author's analysis of 2000 National Election Survey data.

NOTES: Percentages of major party identifiers reflect number of respondents who consider themselves Democrats or Republicans, plus those who consider themselves "closer" to one party. Class categories based on occupation and income: Upper White Collar (25% of total) = White-collar workers with household incomes over $65,000; Lower White Collar (37%) = White-collar under $65,000; Upper Manual (24%) = Blue-collar and service workers over $25,000; Lower Manual (14%) = Blue-collar and service under $25,000.

Table 9.5   Support for Increased Spending on Social Programs by Occupational Class (2000)

	Percent Favoring Increase				
	Schools	Child Care	Social Security	Aid Poor	Welfare
Upper White Collar	75	60	49	44	11
Lower White Collar	79	64	68	53	19
Upper Manual	76	64	70	51	13
Lower Manual	72	67	73	67	27
ALL	76	63	64	52	17

SOURCE: Author's analysis of 2000 National Election Survey.

NOTE: See Table 9.4 for description of occupational class categories.

Class and Political Participation

There is a paradox here. Americans, and working-class Americans in particular, are notably liberal in their attitudes toward spending on education, child care, social security, health care,[2] and help for the poor (as long as it's not called welfare). They are, except for the top class, inclined toward the Democratic Party. But if these things are true, how is it that liberal Democrats are not consistently elected to office and liberal measures are not regularly enacted?

One reason is that the generally more conservative, more Republican people toward the top of the class structure are better informed politically and more likely to participate in political activities than are those toward the bottom. They are, of course, more likely to contribute money to political campaigns. Surveys consistently show that voter participation is higher at higher income levels. There is also some evidence that this class gap in participation is slowly widening (Freeman 2004; Teixeira 1992:Chapter 3). In the November 2004 balloting, for example, people with incomes above the median (around $50,000) were much more likely to vote than those below (see Table 9.6). Thus, middle- to upper–middle-income households are overrepresented in the electorate. Of course, astute politicians can read the numbers, and they tend to craft their message for people above the median. A logical alternative for the Democrats, suggested by sometime presidential candidate Jesse Jackson among others, is to match a liberal platform appealing to working-class interests with a massive effort to increase the electoral participation of lower-income people, both black and white. But many

Table 9.6 Voting in 2004 Election by Family Income

	Percent Voting at Income Level
Less than $10,000	36.5
$10,000–$14,999	39.1
$15,000–$19,999	45.2
$20,000–$29,999	49.4
$30,000–$39,999	54.3
$40,000–$49,999	62.3
$50,000–$74,999	68.1
$75,000–$99,999	74.1
$100,000–$149,999	77.8
$150,000 and over	78.3
Total	60.1

SOURCE: U.S. Census Bureau.

---

[2]According to the General Social Survey, a strong majority of Americans believe we are spending too little on health care (an issue ignored in the NES). Support was slightly weaker among higher-income respondents.

Democratic leaders are reluctant to take this path, fearing that it will send more white, middle-class voters to the Republicans.[3]

Trends in Class Partisanship

Many students of American politics are convinced that social class is a declining influence in American elections. This conclusion is supported by comparisons between "working class" (blue-collar) and "middle class" (white-collar) voters. The gap between these two broadly defined classes in the percentage voting for Democrats has tended to decline over the last half century, especially among white voters (Abramson, Aldrich, and Rohde 1995:152–154; Campbell, Gurin, and Miller 1954). But, as we have noted in previous chapters, the significance of the simple blue-collar/white-collar distinction has eroded in postindustrial society. In both the routinized character of their work and their modest incomes, lower–white-collar workers seem more like blue-collar workers. And the professional occupational category has grown, incorporating many semiprofessionals and technicians in relatively low-status positions.

An analysis by Hout et al. (1995) of voting in presidential elections from 1956 to 1992 showed that both professionals and lower–white-collar workers were, in fact, becoming more Democratic in their preferences, blurring the blue-collar/white-collar line. But they also found that the traditional partisan gap between (Democratic) blue-collar workers and (Republican) mangers or proprietors was widening—proof of the continuing political significance of class defined by occupation.

Partisan differences by household income point in the same direction. In recent decades, lower-income voters have become more likely and higher-income voters less likely to identify as Democrats and vote for Democratic candidates in national elections (Brewer and Stonecash 2007:75–76).[4] As Table 9.7 indicates, the partisan income gap widened significantly in the 1970s, with the shift from the Age of Shared Prosperity to the Age of Growing Inequality, and may reflect the growing awareness of diverging fortunes in the new era.

---

[3]Some students of electoral behavior doubt that nonvoters differ much in their political opinions from voters. See Freeman 2004 and Verba, Schlozman, and Brady 2004. But perhaps the mechanisms that traditionally got reluctant low-income voters to the polls also focused them on what could be considered their own class interests. At the same time, sustained high working-class turnout encouraged politicians to appeal to working-class concerns.

[4]As Brewer and Stonecash show, the trend is robust even among whites considered separately, though the income-class gap is smaller without black voters.

Table 9.7   Party Preference and Household Income, 1952–2006

	Percent Voting for Democrats in U.S. House Elections Household Income		
	Bottom Third	Top Third	Difference
1952–1958	57	49	9
1960s	60	53	7
1970s	66	52	14
1980s	68	51	17
1990s	66	44	22
2000–2006	61	46	15

SOURCES: Brewer and Stonecash 2007:76; CNN.com for 2006 data.

NOTE: Income thirds refer to relative position in the income distribution of *voters*.

Class Conflict and the Labor Movement

The preceding sections focused on electoral politics as an arena of "democratic class struggle." Here, we shift our attention to the conflict between capitalists and workers in the workplace. American labor history has been distinguished by an ironic combination of violent struggle and limited class consciousness that sets the American experience apart from the labor history of other Western nations. Violence has grown out of tenacious capitalist resistance—not so much to specific economic demands as to the very right of workers to organize labor unions. The class consciousness of American workers has been limited in the sense that they and their leaders have typically sought circumscribed goals—basically union recognition, economic security, and decent working conditions, rather than more fundamental changes in the system.

For years, employers exploited differences of race, ethnicity, and skill level among workers—for example, by hiring blacks to replace striking workers—and used violence to suppress strikes. Antiunion violence was common because local and national authorities generally sided with the capitalists. In effect, civil liberties were routinely suspended in strike situations. Without the normal protection of the laws, workers could be physically intimidated (often by thugs hired for this purpose), union organizers harassed, and leaders jailed. Much of this activity was coordinated by special firms that sold "union-busting" services (Litwack 1962: 95–115).

In Chapter 3, we sketched the history of the labor movement to World War I. A brief review of the decades that followed will provide a framework for understanding recent developments (Boyer and Morais 1975; T. Brooks 1971). Although unions expanded during the conflict, a postwar "red scare" and determined employer resistance cut membership from 20 to 10% in the 1920s (Brooks 1971:148). Change came with the Great Depression and the concomitant shift in national politics, particularly during the years 1933 to 1937. A labor historian has described this period as "the highwater mark of class struggle in modern American history" (M. Davis 1980:47). When the United Textile Workers announced an industrywide strike in 1934, *Fibre and Fabric*, the New England trade journal, declared, "A few hundred funerals will have a quieting influence" (*Fortune* 1937:122). Before this bitter, violent strike had completed its 3-week run, thousands of National Guard troops had been mobilized in seven states, and 12 strikers and one deputy had been killed. The union lost.

In this period, however, there were more labor victories than defeats. Some of the most significant were gained through a new tactic—the sit-down strike: workers forced concessions from employers by taking physical control of the workplace. First used in the rubber industry in 1936, the innovation, which appealed to the militant mood of workers at the time, spread rapidly. One labor official remembers 1937 as the year he received calls daily like the one from a drugstore food-counter worker: "My name is Mary Jones; I'm a soda jerk at Liggett's; we've thrown the manager out, and we've got the keys. What do we do now?" (Brooks 1971:180).

By 1938, the right to union representation had been written into law through the National Labor Relations Act, known as the Wagner Act, and unions had successfully established themselves in the mass-production industries, such as steel, automobiles, rubber, and electrical goods, which stood at the center of the American economy. A conjunction of social and political developments made these accomplishments possible. Of critical importance was overcoming the division within the working class and the labor movement that had plagued earlier unionization drives. The significance of ethnic differences declined as the sons and daughters of immigrants joined the labor force. No longer cut off from one another by language barriers, more confident of their place in American society than their parents had been, and more demanding of their rights, these second-generation Americans helped recast labor relations just as they contributed to the revamping of national electoral politics.

Differences between skilled and unskilled or semiskilled workers in manufacturing did not disappear, but a barrier to unionization was removed when the Committee for Industrial Organization (CIO) was formed within the old American Federation of Labor (AFL) in 1935, with the explicit purpose of organizing workers on an industry-by-industry basis rather than on the craft basis that was typical of the AFL unions. The following year, the more aggressive CIO broke with the tradition-bound AFL, retitling itself the Congress of Industrial Organizations. The CIO strove to remove another source of weakness by organizing both black and white workers.

As CIO organizers set about their task, the American working class was in an extraordinarily militant mood. The story of Mary the drugstore striker may be apocryphal, but it suggests the atmosphere of the times. The rank and file frequently ran ahead of union organizers, who found themselves forced to restrain premature action they were not in a position to support. Working-class solidarity grew to the extent that big strikes attracted workers from other industries and localities, who came to offer moral and even physical support (Greenstone 1977:44).

A key to the worker militancy of the 1930s was the experience of the Depression. Since Marx, social scientists have recognized that economic insecurity feeds class consciousness. The insecurity that workers experienced during this period was connected to the breakdown of the entire economic system. The confidence that workers had in their employers was shattered by the recognition that even such powerful companies as U.S. Steel and General Motors were subject to the vagaries of the marketplace and apparently indifferent to the fate of their workers. When capitalists responded to the Depression by laying off workers, reducing benefits, and speeding up work, they lost the loyalty of many workers.

But neither the reduced factionalization nor the growing militancy of the working class could have accomplished the transformation of industrial relations of the 1930s without the changed political context represented by the liberal New Deal, and the passage in 1935 of the Wagner Act, which guaranteed the right of workers to form labor unions, prohibited employers from interfering with the exercise of that right, and set up a National Labor Relations Board (NLRB) to ensure compliance. As the experience of the 1920s made clear, if government was hostile to labor, or at least so indifferent as to ignore patently illegal forms of employer resistance, unionization was impossible.

The Postwar Armistice: Unions in the Age of Shared Prosperity

If the 1930s represented a period of explicit class conflict during which these changes were forced on the capitalist class, the decade after World War II was the time when the details of a class armistice were worked out. When leaders of the major labor organizations met in Washington with key business representatives at the end of the war, their purpose was to "lay the basis for peace with justice on the home front." President Truman convened the conference at the suggestion of conservative Senator Arthur Vandenberg, who told the president, "Responsible management knows that free collective bargaining is here to stay . . . and that it must be wholeheartedly accepted" (Brody 1980: 175).

Although many business leaders were willing to accept the existence of unions, they were determined to preserve for the capitalist class what they termed the "right to manage." At stake was participation in decisions regarding such matters as investment (including plant openings and closings), product design and production methods, and the pricing of final products. Had labor gained a share in these decisions, as did some contemporary European unionists, the labor movement could have had meaningful influence on employment and other basic economic questions. But management insisted on the notion of "property rights," refusing to even concede to unions the right to examine the books of the enterprises with which they bargained. The right to manage gradually ceased to be an issue (Brody 1980: 173–213).

By the late 1950s, the shape of the industrial peace was unmistakable. Unions were firmly established among blue-collar workers at the core of the economy in heavy industry. Here they could gain substantial benefits for their members as long as they did not interfere with management prerogatives—benefits that would allow a large segment of the working class a life of relative affluence. "The labor movement," concluded auto union leader Walter Reuther, "is developing a whole new middle class" (Brody 1980:192). Serious industrial conflict was banished to the periphery of the economy—to the smaller firms, weaker economic sectors, and backward regions (especially the South). If such conflict was relatively infrequent, that was mostly because the labor movement had grown satisfied and unaggressive. The giant AFL-CIO (the two had re-merged in 1955) was behaving, in the words of labor economist Richard Lester, like a "sleepy monopoly" (Dubofsky 1980:8).

Ironically, labor came to play a more dynamic role in national electoral and legislative politics than it did in the workplace. Long reluctant to involve itself directly in politics, the labor movement had become a major supporter of the Democratic Party and a broad array of liberal social and economic programs. Labor now spoke not just for union members but also for the working and lower classes generally. Unions were politically active in two broad arenas: electoral and legislative. In the former, the labor movement promoted liberal candidates, both within and on behalf of the party; raised a substantial part of the party's campaign money; and fielded thousands of campaign workers. In areas such as Detroit, where unions were especially strong, the party and the union's political organization became virtually indistinguishable (Greenstone 1977:119–140). In Washington, labor maintained a formidable lobbying apparatus.

During the 1960s and early 1970s, the labor lobby played a major role in obtaining passage of liberal legislation in such areas as civil rights, health care, minimum-wage protection, public employment programs, nutrition programs for the poor, and occupational health and safety (Greenstone 1977). During these years, union lobbyists were frequently the leaders of broad liberal coalitions that confronted business lobbies and other conservative groups over critical pieces of legislation. In effect, a class cleavage ran through the center of national legislative politics, and the unions were critical players on one side. Although that cleavage disappeared from view over many issues (such as the Vietnam War, abortion, and gun control), it was still visible during the 1990s and early 2000s in the struggles over budget and tax policies and issues such as the minimum wage, parental leave, and health care.

Labor in Decline

The Age of Growing Inequality has been a period of devastating decline for the American labor movement. The clearest measure of labor's fate was the drop in the proportion of workers who belonged to unions. Membership rates, which had been slowly eroding since the mid-1950s, plunged in the 1980s. By 2004, only 12.5% of workers were union members. And this meager figure, bolstered by higher rates among public sector workers, masked the labor's weakness in the private sector, where under 8% were union members. The position of the American labor movement had not been so weak since the 1920s.

Labor's problems stemmed to some extent from basic shifts in the economy and occupational structure, which we discussed in Chapter 3. Manufacturing employment, the bastion of union strength since the 1930s, was in decline in the 1980s. Facing increased competition from abroad and shrinking profits, American industrial companies were closing older U.S. plants and moving operations to low-wage havens abroad or to the antiunion states of the Sunbelt. Although manufacturing output grew during the decade, one of every three jobs in heavy industry disappeared (Berman 1991:7).

Employment growth in the Age of Growing Inequality was strongest among those categories of workers who are the most difficult to organize: white-collar employees, service workers, employees of small establishments, and female workers. (Most of these groups are among those we have found to be the least class conscious.) The only area in which the unions made significant gains was in the public sector, whose future employment growth is uncertain at best. In short, labor's natural economic base was crumbling.

Labor's organizational decline was soon reflected in political weakness. The 1978 battle over the Labor Reform Bill in Congress gave an early warning of what was to come. Both labor and business regarded the measure, which sought to restore the effectiveness of the 1935 Wagner Act in protecting workers' rights, as a critical test of strength. The legislation was critical to union hopes of regaining lost ground. Over the years, employers had discovered that they could stave off union organizing efforts by legal maneuvering and other tactics designed to discourage union activists and postpone representation elections. The longer the delays stretched out, the greater the probability that a union drive would collapse. The basic provisions of the bill were designed to guarantee speedy elections to determine whether workers wanted union representation, to ensure prompt decisions from the NLRB in unfair labor practice cases, and to stiffen penalties for violation of existing labor laws, such as the firing of union sympathizers.

The AFL–CIO spent $3 million on a hard-fought campaign in support of the bill. A coalition of business groups, including the Business Roundtable, National Association of Manufacturers (NAM), U.S. Chamber of Commerce, and National Federation of Independent Businessmen, spent approximately $5 million to secure its defeat (Cameron 1978:80).

Unionists were stunned by the defeat and the composition of the coalition that had opposed them. They were used to the idea that the lesser capitalists represented by the Independent Businessmen, the Chamber of Commerce, and the NAM harbored strong antiunion sentiments. But the Business Roundtable represented the largest American corporations, the firms that had made their peace with the unions in the 1950s. The Roundtable's role in thwarting the bill came as a bitter revelation. Shortly after the bill was stopped by a filibuster in the Senate, Douglas Fraser (1978), president of the United Auto Workers, resigned from the semiofficial Labor–Management Group in Washington with a direct attack on his Business Roundtable colleagues in the group:

I believe leaders of the business community, with few exceptions, have chosen to wage a one-sided class war today in this country. . . . The leaders of industry, commerce, and finance in the United States have broken and discarded the fragile, unwritten compact previously existing during a past period of growth and progress.

Ronald Reagan, elected in 1980, proved to be the most antiunion president in decades. Reagan signaled his intentions early in his presidency by firing thousands of striking air traffic controllers and putting their union out of business. Under his administration, the protections of labor law were further diluted by pro-management appointments to the NLRB and the Labor Department. One experienced union lawyer commented that the labor board had come to operate "like a bloodless bureaucratic death squad" (Geoghegan 1991, quoted in Berman 1991:7). In the 1978 elections, for the first time, business PACs outspent labor PACs (*New York Times*, Aug. 4, 1981). In Congress, the ability of labor lobbyists to move (or block) legislation was waning. In a period of racing inflation, for example, they could not obtain a raise in the minimum wage.

Business, encouraged by the new atmosphere in Washington and threatened by foreign competition, assumed a more demanding stance in labor relations. Many firms forced unions to yield "givebacks" of favorable wage rates and working conditions won in earlier negotiations. Others sought to destroy their unions by forcing decertification elections or provoking a strike and hiring nonunion replacement workers. In the Sunbelt, corporations found a legal and political atmosphere in which union organizing was extremely difficult.

Even large, established corporations were taking advantage of the services of a new breed of labor consulting firms. Eschewing the violent union-busting methods of the past, the new consultants were masters of manipulation, sophisticated in their application of social-scientific knowledge. For example, the consultants showed management how to convey to employees an artificial sense of participation in company decisions and how to use systems of subtle rewards and punishments to influence employee attitudes or create anxieties that undercut pro-union sentiment. Employers were advised to avoid certain categories of workers, because they are more likely to be receptive to union appeals. Consultants also instructed corporations in the advantages they could glean from calculated violations of poorly enforced labor laws (Langerfeld 1981).

Organized labor contributed to its own problems. After the purges of leftist activists in the late 1940s, the well-paid officers of many unions had grown complacent, in a few cases corrupt, and distant from the problems of ordinary workers. Such leaders might have been qualified to conduct the daily business of established unions in quiet times, but they were less effective against the swirling currents of political and economic change.

Labor's defeats in the Age of Growing Inequality have real implications for the dynamics of the American class system. The unions always spoke for a much wider constituency than their own members. In the workplace and in national politics, however indirectly and imperfectly, organized labor has represented the interests of working-class Americans against those of the capitalist class. At work, the unions set a standard that even the employers of nonunion workers had to acknowledge. In politics, union voters, activists, and campaign contributions backed liberal candidates at all levels. Labor's long decline, along with the rise of business power, has altered the balance of power among social classes in ways that may affect this country for years to come.

Conclusion

Our concerns in this chapter have been class consciousness and class conflict. Marx first raised the issue of class consciousness because he believed that subjective awareness of one's objective position in the property and occupational system would lead to a sense of belonging with others of similar position and promote conflict with those above or below. Marx's conception assumes the causal sequence we outlined at the beginning of this chapter: objective class position leads to class consciousness, which in turn shapes political behavior. Under what conditions, Marx asked, is class consciousness likely to emerge? He emphasized the importance of the following factors: concentration and communication, absolute or relative deprivation, economic insecurity, alienation at work, class polarization, homogenization of the proletariat, and political organization and struggle.

As these factors suggest, Marx's thinking about class consciousness focused on societywide developments over extended historical periods. Modern social science researchers, like Richard Centers, have tended to focus on individual opinion measured at a point in time. Centers got at one aspect of class consciousness with a simple survey question about class identity. He showed that (in response to a force-choice question) the vast majority of manual workers call themselves working class and most nonmanual workers consider themselves middle class; very few Americans would brand themselves upper or lower class. These class labels carry connotations much broader than just a classification of jobs. They imply a way of life, a set of friends, and a system of values. And, they cover all members of a family, not just the principal breadwinner, whose job is usually used to define class membership.

In the Western democratic countries, class consciousness has not expressed itself in the sort of revolutionary upheaval that Marx anticipated. Instead, class conflict has been channeled into electoral competition and labor politics. The democracies typically have a right-left spectrum of parties, with the left parties tending to draw their members from the bottom of the class structure and the right parties from among the privileged.

The United States fits this general pattern, but the class identities of our major parties have always been somewhat blurred by ethnicity, religion, and regional loyalties, and the class gap in party support has never been as great as the corresponding difference in European systems. From the 1930s until the 1970s, the Democratic Party dominated national politics. The party was built on a working-class base, reinforced by traditional ethnic and regional loyalties. In recent years, the strength of this so-called New Deal (or Democratic) coalition has eroded, changing the shape of national politics.

National surveys show that Americans are still influenced by class position in their political opinions and electoral preferences. But national politics has taken a more conservative turn than might be predicted from this national polling data. There is no simple explanation for this shift. Among the factors that appear important are (1) the effect of superior rates of political participation (including voting) at higher class levels; (2) the growing power of money in American politics described in Chapter 8; (3) the influence of divisive issues such as affirmative action, welfare, abortion, and gun control; and (4) the waning strength of labor unions and urban political machines.

The American labor movement consolidated its position at the same time that the New Deal coalition emerged under Roosevelt in the early 1930s. Labor became one of the pillars of the Democratic Party. Throughout the 1930s, American labor history was marked by violence and management resistance to the right of workers to unionize. The post–World War II period produced an "armistice" in labor relations, as the major corporations accepted unions as a fact of life. But the armistice came undone in the late 1970s and 1980s—a characteristic development of the Age of Growing Inequality. Business, often supported by conservative Republican administrations in Washington, became much more assertive in its confrontations with labor. Union strength declined sharply, contributing in turn to the relative decline of the Democratic Party.

These developments, along with the increased political activism of corporations and the capitalist class described in the last chapter, have produced a critical shift in the class balance in American politics in favor of the capitalist class.

Key Terms Defined in the Glossary

alienation
class consciousness
class identification

class position, objective
political action committee
(PAC)

Suggested Readings

Brewer, Mark D. and Jeffrey M. Stonecash. 2007. *Split: Class and Cultural Divides in American Politics*. Washington, DC: CQ Press.
  Argues that both class and culture influence voters and politicians.

Brody, David. 1993. *Workers in Industrial America: Essays on the 20th Century Struggle*. 2nd ed. New York: Oxford University Press.
  A lively introduction to the history and historiographic literature of the labor movement in the twentieth century.

Edsall, Thomas and Mary Edsall. 1991. *Chain Reaction: The Impact of Race, Rights, and Taxes on American Politics*. New York: Norton.
  The "wedge" issues that splintered the Democratic class–ethnic coalition.

Geoghegan, Thomas. 1991. *Which Side Are You On: Trying to Be for Labor When It's Flat on Its Back*. New York: Farrar, Straus & Giroux.
  A labor lawyer's personal account of the 1970s and 1980s.

Manza, Jeff and Clem Brooks. 1999. *Social Cleavages and Political Change: Voter Alignments and U.S. Party Coalitions.* New York: Oxford University Press.

  Social factors, including class, race, gender, and religion, in U.S. presidential elections since 1948.

McCarty, Nolan, Keith T. Poole, and Howard Rosenthal. 2006. Polarized America: The Dance of Ideology and Unequal Riches. Cambridge: MIT Press.

  The relationship between growing inequality and polarization in American politics.

Rieder, Jonathan. 1985. *Canarsie: The Jews and Italians of Brooklyn Against Liberalism.* Cambridge, MA: Harvard University Press.

  Political ethnography. The disintegration of the Democratic coalition from the bottom up.

Teixeira, Ruy and Joel Rogers. 2000. *America's Forgotten Majority: Why the White Working Class Still Matters.* New York: Basic Books.

  The majority of voters are white men and women, without college degrees, in low-status jobs, many of them victims of the transformed economy. The Democratic Party should appeal to this working-class majority, say the authors.

Thompson, E. P. 1963. *The Making of the English Working Class.* New York: Vintage.

  Classic historical portrayal of the development of working-class consciousness.

CHAPTER 10

The Poor, the Underclass, and Public Policy

We stand at the edge of the greatest era in the life of any nation. . . . Even the greatest of all past civilizations existed on the exploitation of the misery of the many. This nation, this people, this generation, has man's first chance to create a Great Society; a society of success without squalor, beauty without barrenness, works of genius without the wretchedness of poverty.

Lyndon B. Johnson (1964)

[T]his legislation will end welfare as we know it.

Bill Clinton (1996)

D uring the era we have titled the Age of Growing Inequality, Americans added a new word to their political lexicon: homelessness. The term was not a recent addition to the dictionary, but beginning in the 1980s, it was being used in a new way—to name a social problem. And although homelessness had once referred to people without fixed residences, who drifted from place to place, it was now being applied to people who literally had no access to conventional housing, hundreds of thousands of people who slept in doorways, packing crates, bus stations, and shelters. A disturbing aspect of this phenomenon was the large number of families with children among the homeless population. (One-third of those who used a homeless shelter during a 3-month period of 2005 were members of families with children.)

According to an official estimate, there were approximately 750,000 homeless people on any single day in 2005. But a much larger population, 1.5 to 3.0 million, were likely to have an episode of homelessness at some point in the course of a year. And because the poor of 2005 were poorer in an absolute sense than the poor of the early 1970s and the real cost of housing was higher, there was a growing population that was vulnerable. Nearly 9 million families were spending over half their income on housing, a level that often leads to homelessness.[1]

Visible on city streets, the homeless population was the ominous tip of the poverty iceberg. In 1980s and early 1990s, poverty rates rose to levels not seen since the 1960s. By the official count, from 1981 to 2005, there were never fewer than 30 million poor Americans. Rising childhood poverty rates were accompanied by increased hunger among children and their families.[2]

Most Americans and their leaders in Washington gave little thought to the problem of poverty. They were more concerned about the performance of the national economy and their own (in many cases, shrinking or stagnant) incomes. Americans were not as confident as they had once been in the capacity of government to tackle big social problems. Many were convinced that the poor were responsible for their own situations. The problem, according to conservatives, was not poverty but "welfare dependency." As we will see, the national will to fight poverty has waxed and waned in recent American history.

In this chapter, we deal with several key questions: How is poverty defined and measured? How many Americans are poor? Who are the poor? What are the long-term trends in poverty? What are the causes of poverty? How has government

---

[1]*New York Times* 2001; Rossi 1989; U.S. Dept. of Housing and Urban Development 2007; Wright 1989.

[2]The nonpartisan National Commission on Children in its 1991 report to the president and the Congress noted that estimates of childhood hunger ranged from 2 to 5.5 million but "there is no doubt that the problem has increased over the past decade" (p. 124). A 2002 Department of Agriculture survey classified 17% of households with children as "food insecure"—that is, they were "uncertain of having, or unable to acquire, enough food for all their members" at some time during the year. More than 9% of families with children reported that they could not feed their children balanced meals, 4% reported that children were not eating enough, and 1% that children were "hungry" (Nord et al. 2003).

policy responded to poverty? We will begin by going back to the 1930s, when poverty was first recognized as a problem requiring federal action, and the modern system of social programs was born under Franklin Roosevelt's New Deal.

The Beginnings of Welfare: Roosevelt

The Great Depression hit the nation with devastating effect. The unemployment rate was above 20% from 1932 through 1935 and did not go below 15% until the eve of war in 1940. Those who had jobs saw their wages fall. Frightened, angry citizens joined widespread and often violent protests. The efforts of private charities and local governments to provide assistance were overwhelmed by the magnitude of the crisis. After the election of Franklin D. Roosevelt in 1932, for the first time, the federal government assumed primary responsibility for providing assistance in an economic crisis.

The Roosevelt administration moved quickly and devised entirely new approaches to the problems of unemployment and poverty. The federal government began a massive program of "direct relief"—cash payments to families in desperate need, managed by the states. Like many Americans, the administration was uncomfortable with the idea of direct relief, except as an emergency measure. Emphasis soon shifted to job relief. Under the federally funded Works Progress Administration, millions of workers were hired, often to work on needed public works projects, such as roads, bridges, and schools.

The main responsibility for direct relief was turned back to the states and localities, but with the federal government sharing the costs of supporting certain categories of people considered unemployable: the blind, the physically or mentally disabled, the elderly, and mothers who had small children but no husbands to provide for them. (The small program for mothers eventually grew into Aid to Families with Dependent Children [AFDC].) The programs created in response to the emergency of the Depression were soon phased out, but a new federal program was created to provide workers and their families a long-term safety net.

The core of that new system was the 1935 Social Security Act, which established a national social insurance system. Enrolled individuals (not everyone was included, especially in the earlier years) would receive full coverage after 10 years of contributions made by employees and employers to the trust funds. Retired persons would get permanent pensions, as would those who were disabled to the point of not being able to work. Widows and orphans (survivors) of insured workers would get benefits. A system of unemployment insurance was also established that would give temporary payments to insured workers during periods of layoff, usually up to 26 weeks. Unemployment benefits not only protected workers' families, but also provided a safety net for the economy by maintaining consumer purchasing power during slack periods.

These programs, and others that would be added later, came to be called "entitlements," because all who meet specified prerequisites are entitled to receive them and the government is committed to providing the necessary funding. Some entitlements, such as AFDC and related public assistance programs, are "means-tested": they are only available to people with incomes below a specified threshold. Others, such as old age benefits under Social Security, are available to people of all income levels.

These features would, in the long run, have significant political consequences. Entitlement spending, because of its open-ended character, proved difficult to contain. When mounting budget deficits became a national concern in the 1990s, means-tested entitlements were the natural target because of their narrower and relatively powerless constituency—the poor.

Rediscovery of Poverty: Kennedy and Johnson

During the Age of Shared Prosperity after World War II, the scene had changed. The Social Security system was beginning to pay out large sums to the elderly, and because it was viewed by the public as an insurance program that returned to them in the form of pensions—money that they had earlier contributed in the form of taxes on wages—it was not stigmatized as "relief" and was popular among all segments of society. (Actually, the program taxes current workers to pay retirees, who generally receive much more than they pay in, but most people do not think of it that way.) The "make-work," or special public jobs, had disappeared. The unemployment compensation system was working smoothly and was taken for granted.

Although these programs were created by New Deal Democrats over Republican opposition, a national, bipartisan consensus developed around them. When the Republicans regained the White House in the 1950s, they did not attempt to undo what the New Deal had done. A benign mood had settled on the generally prosperous country, and neither party had much taste for innovations in social policy.

Around 1960, two influential books challenged the nation's complacent mood: John Kenneth Galbraith's ironically titled *The Affluent Society* (1958) and Michael Harrington's *The Other America: Poverty in the United States* (1962). Galbraith and Harrington reminded the country that many Americans, perhaps a fourth of the population, remained poor in the midst of general prosperity.

The new generation of Democrats who came to Washington with John Kennedy in 1961 was keenly aware of these problems. The president and his influential brother Robert had been shocked by the misery they saw in West Virginia during the campaign. They and many other members of the administration had read Harrington's book. Galbraith became an advisor to the president.

The Kennedy and Johnson (1961–1968) administrations would make important improvements in both the welfare and social security systems. They added three new programs: the food stamp program, which reduced malnutrition (and simultaneously helped farmers) by providing low-income families with scrip redeemable for food at retail stores; Medicare, guaranteeing health insurance for the elderly; and Medicaid, providing health insurance for the poor, especially children on public assistance and (increasingly) seniors receiving long-term care in nursing homes.

This system of social protections was broadened under Johnson's successor, Richard Nixon, with Supplemental Security Income, which provides income supplements for needy elderly, blind, and disabled people; a substantial expansion of the food stamp program, undertaken in the wake of revelations of widespread hunger among the poor in America; and crucial legislation guaranteeing that Social Security payments would rise with the cost of living.

Lyndon Johnson regarded what he called the "War on Poverty" as one of the top priorities of his administration. In a 1965 speech, Johnson proclaimed, "We stand at the edge of the greatest era in the life of any nation. For the first time in human history, we have the abundance and the ability to free every man from hopeless want, and to free every person to find fulfillment in the works of his mind or the labor of his hands." From the cold perspective of the early 2000s, Johnson's grandiloquent rhetoric sounds quaint, and his optimism appears misplaced. In fact, the United States made steady progress against poverty in the 1960s and early 1970s, but little if any thereafter. More serious, Americans lost the hope and commitment that Johnson and many of his contemporaries displayed.

The Official Definition of Poverty

Officials in the Kennedy administration quickly recognized that to accurately assess the problems of the poor and develop remedial measures, they needed a reliable official definition of poverty—something the government had never before enunciated. Using the definition, the government could measure the number and characteristics of the poor and analyze the causes of their poverty, and later gauge the effectiveness of antipoverty measures.

The poverty standard finally adopted by the administration was designed by Mollie Orshansky (1974), an economist at the Social Security Administration. Orshansky put two pieces of information together from government surveys: the cost of a minimum nutritious diet for a typical family of four and the proportion of income (approximately one-third) that the average family then spent on food. Multiplying the price of the food budget by 3 to allow for nonfood costs, Orshansky calculated an income "poverty line" of approximately $3,000 for a family of four. If you were a member of a four-person family with an annual income below $3,000, you were poor by this standard. On the assumption that family needs vary with size, somewhat higher and lower poverty thresholds were computed, on a similar basis, for households that were larger or smaller than the typical four.

Orshansky's poverty line became the official federal standard. Each year, the Census Bureau uses it in conjunction with the annual survey of household income to estimate the number of poor people in the country. Any family whose total pre-tax income is lower than the poverty threshold for its size is counted as poor. By this new standard, 22% of Americans were poor in 1960.

Because prices change, the poverty line must be regularly adjusted for inflation. At first, this was done by sending government employees to grocery stores to determine the cost of feeding a family of four, the basis of the poverty standard. Later, poverty thresholds were simply adjusted each year in proportion to the annual increase in the Consumer Price Index, the government's general measure of inflation. By 2005, the poverty line for a family of four was about $20,000.

Note that the poverty standard is adjusted for changes in prices ($20,000 was required in 2005 to purchase what $3,000 bought in 1960) but not for changes in the general standard of living. It does not take into account the fact that, on average, Americans lived at a higher material standard in 2000 than they had in 1960. By recycling this standard year after year, the government is saying something about the meaning of poverty. It is telling us that what defines people as poor is their material deprivation in an absolute sense, rather than the relative gap between their standard of living and the standard typical of people in the same society. We will come back to this question shortly.

The Orshansky standard was as reasonable a poverty measure as anyone could come up with at the time. But it was problematic from the beginning and remains a target of controversy to this day. The trouble started when the proposed measure was being considered by Kennedy's Council of Economic Advisors. Government nutritionists had come up with not one, but two family food plans. (Both were based on the dubious assumption that the family cook was a sophisticated dietitian who never wasted a penny.) The first food plan was an emergency diet, suitable to maintain a family for a short period. The second plan, which cost 25% more, was designed to provide the nutrition necessary for long-term family health. It was left to the Council of Economic Advisors to decide which food plan to use as the basis for measuring poverty. Adopting a standard based on the second plan would result in a higher poverty line and, as it turned out, a much higher estimate of the number of impoverished Americans. That was apparently unacceptable to the Kennedy administration. The Council of Economic Advisors chose to base the official poverty line on the emergency food plan, thus opting for a more restrictive definition of poverty. The decision had little to do with nutrition and nothing to do with economics; it was essentially political.

The higher standard, now referred to as "125% of the poverty line," is used in some government statistical reports as a broader measure of poverty. The population that falls between the official poverty line and 125% of the poverty line includes many of those who belong to the class we call the working poor.

In principle, the poverty line was based on an objective, scientific standard: minimum nutritional need, plus a proportional allowance for rent and other essentials. In retrospect, however, it appears that there is no wholly objective way to specify minimum requirements. Even when political considerations don't get in the way, contemporary cultural values always intercede and help define what is appropriate. When the general standard of living goes up, so do ideas about the minimum needed to maintain subsistence. If there was any doubt about this, it was removed by the research of Oscar Ornati (1966) and Rainwater (1974).

Ornati collected the subsistence budgets that had been used by local social service agencies, going back to 1905. He found that these budgets, which were designed to provide temporary relief for family survival, were typically based on the same notion of a "basic-diet-plus" that was adopted for the federal poverty standard. Ornati discovered that the conception of a subsistence budget tended to expand as national living standards improved. For instance, a 1908 budget for a family of four included 22 pounds of high-protein foods. A 1960 budget assumed that a family of this size needed 55 pounds. The 1908 budget makers supposed that a family needed four rooms but did not require a bath; by 1960, the standard was five rooms and a full bath. Sociologist Lee Rainwater (1974) subsequently showed that the cost of the budgets Ornati analyzed was consistently close to 40% of the current average disposable income for a four-person family. By no small coincidence, the same was true of the federal standard when it was adopted in the early 1960s. Rainwater also showed that popular ideas about what a family needs to "get along," as reflected in successive Gallup polls, were always around 50% of the average family disposable income for the year the question was asked.

The conclusion that can be drawn from the research of Ornati and Rainwater is that conceptions of a minimum living standard are inevitably relative. They depend on income levels and lifestyles in the societies where they are made. This generalization apparently applies with as much force to the experts who design poverty standards as it does to popular opinion. Thus, what we now consider a poverty-line existence would have qualified as middle-class comfort at the beginning of the last century and might still be regarded as such in some small town in the Peruvian Andes.

This brings us back to the federal standard. If we could somehow summon the Kennedy administration experts, erase their collective memory of the poverty line they created in the 1960s, and make them do it again in our own time, they would almost certainly come up with something that reflected current living standards. But for thirty years, the government has treated the poverty line as an absolute standard, one that is annually adjusted for increasing prices, but not for changing lifestyles.

Most sociologists believe that the government should measure poverty with a relative standard that would change with the changes in the average standard of living. They reason that what makes people think of themselves as poor and causes others to regard them as poor is the comparison between their lives and the mainstream lifestyle in their community. The relative standard most frequently proposed is half the median family income (that is, a typical family with income lower than half the median income would be considered poor). In 1960, the median family income was around $6,000, which meant that the official poverty line of $3,000 for a four-person family was, in fact, close to half the median income when it was first minted. By 2005, the official standard, adjusted for inflation, had fallen to 35% of the median family income—or, more precisely, the median had increased, leaving the government poverty line behind.

The official poverty standard has other defects. It is based on *money income* as measured by the government's annual household survey. The survey's money income figures do not reflect taxes, either the taxes that the poor pay or the substantial payments many now receive through the Earned Income Tax Credit (more about the EITC later). And money income does not include the noncash (so-called "in-kind") benefits given to the poor in such forms as food stamps, subsidized housing, and health care. At least some of these are the equivalent of money, and including their value in family income would lower our estimates of poverty.

Most analysts would agree that the federal poverty measure is in need of an overhaul, though agreement about exactly how it should be changed is more elusive.

How Many Poor?

Somewhere between 37 and 54 million Americans are poor, according to the estimates in Table 10.1. The first two are the official statistic and "125% of the poverty line," both produced by the Census Bureau and based originally on Orshansky's minimum family budgets. The third is a revised version of the official standard, based on recommendations from a National Academy of Sciences panel of experts, that takes into consideration the value of in-kind benefits, taxes paid, the EITC, work expenses like child care, and other factors. Remarkably, the multiple adjustments, at least for the year shown here, seem to have cancelled one another out and produced an estimate close to the official starting point. These three estimates are based on absolute definitions of poverty. The last and highest estimate is based on the most commonly proposed relative standard, half the median family income. At 54 million people, it gives us an idea of how many poor we would find when using a standard that bears the same relationship to mainstream living standards as the official one did when adopted in 1960.

So how many poor are there? An easy answer is 44 million, the average of the four measures in the table. But, for reasons described in the next section, we generally use statistics based on the official standard for the remainder of this chapter.

Who Are the Poor?

The debate about how to measure poverty is endless, and it is not difficult to imagine why. Any serious discussion of poverty in an affluent society, which also regards itself as democratic, inevitably stirs political emotions. Just below the surface, the important technical dispute about measurement is also a debate about the fairness of our political and economic institutions.

To examine poverty, we will have to settle on some measure, and the official standard is the best choice for now. Despite its defects as a measure of total poverty, the official standard is fairly reliable for comparing subgroups of the population or tracking change over time. It happens also to be the basis of most of the detailed poverty statistics published by the government. For the remainder of the chapter, all our statistics, unless otherwise noted, will be Census Bureau numbers based on the official standard.

Table 10.1   Four Estimates of Poverty in the United States, 2004

Standard	Poverty Rate in Percent	Millions in Poverty
Official	12.7	37
125% Official	17.1	49
NAS Alternative	12.5	36
Relative Standard	18.5	54

SOURCES: U.S. Census Bureau; Mishel et al. 2007: 294, 201.

Figure 10.1 uses government statistics for 2005 to answer a key question: who are the poor? It starts with the total number of poor people, 37 million, and breaks this figure down in various ways. Here are some of the things we learn:

- There are many more white poor than black or Hispanic poor.
- A high proportion of the poor are children. Few are elderly.
- There are about as many poor in families headed by married couples as in families headed by single women. (Families, poor or otherwise, headed by single men are still relatively rare.)
- Only a minority of the poor live in the central cities of metropolitan areas. The majority are spread out among suburbs, small towns, and rural areas.

There is enough information here to contradict some popular stereotypes. In particular, the typical poor person is obviously not a member of a black, female-headed family, living in a central city. As it turns out, less than 1 in 10 of the poor fit that description.

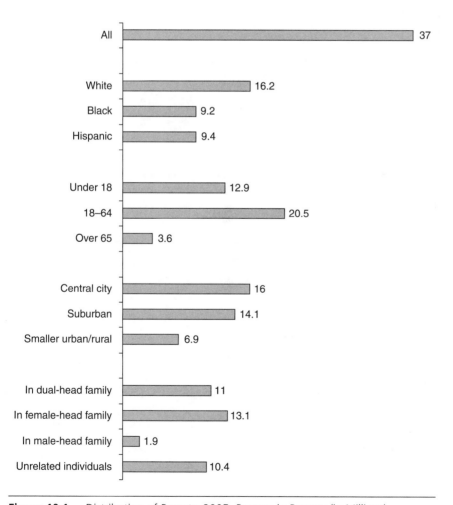

Figure 10.1   Distribution of Poverty, 2005. Persons in Poverty (in Millions)

Figure 10.2 answers a different kind of question: if you belong to a certain social group, what are your *chances* of being poor? This life chance is called the risk of poverty (or poverty rate). It is simply the percentage of the people in a group that falls below the poverty line. Just a quick glance at the graph reveals that there are very large differences in the risk of poverty. The overall rate is 12.6%. Blacks, although they are a minority of the poor, are three times more likely than whites to be poor. Children are at a much greater risk of poverty than adults. Approximately 30% of black and Hispanic children are poor. And whatever their ethnicity, female-headed families have poverty rates far above those of families headed by couples.

As we noted earlier, the measure of poverty employed here counts all money income—including, of course, the transfer payments people receive from the government, including public assistance, social security, veterans benefits, and unemployment. These payments make a big difference. Without them, the poverty rates would be much higher. But the effect of cash transfers is uneven, as Table 10.2 demonstrates. Note that seniors and female-headed families with children have

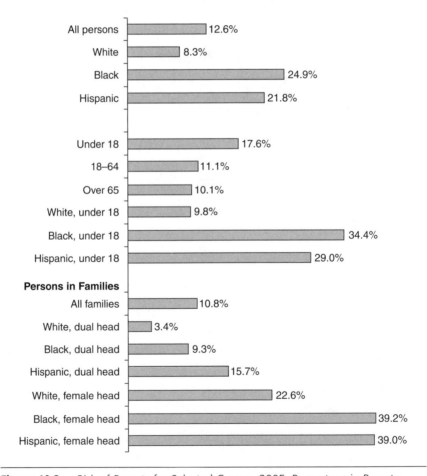

Figure 10.2   Risk of Poverty for Selected Groups, 2005. Percentage in Poverty

virtually the same high pretransfer poverty rate. But a dramatic gap opens between them when the cash transfers are included. The poverty rate of children is hardly touched by transfers. Whatever advantages poor children receive from cash programs, they are clearly not intended to lift families out of poverty. The one government transfer program that significantly reduces poverty is social security for the elderly. The success of that program is not unrelated to the fact that the elderly are well organized in political pressure groups and vote in substantial numbers, while the adult relatives of impoverished children are an unorganized population with low rates of political participation.

Table 10.2   Poverty Rates Before and After Cash Transfers, 2000

	Before Cash Transfers	After Cash Transfers (Official Rate)
All persons	18.6	11.3
Under 18	18.7	16.2
Over 65	48.2	10.2
Female head with children	40.9	35.1

SOURCE: U.S. Census Bureau.

Trends in Poverty

When Lyndon Johnson became president and promised to build a society free from poverty, ignorance, and exploitation, the poverty rate was already falling. As Figure 10.3 shows, the rate fell steeply through the 1960s, more or less flattened out in the mid-1970s, and actually climbed in the 1980s and early 1990s. Since then, it has fluctuated, but never dipped beneath the low point it reached in 1973.[3] Like the trends we observed for wages, income, and wealth, the path of poverty rates shifted with the transition from the Age of Shared Prosperity to the Age of Growing Inequality.

Over the years, as the poverty rate has fluctuated, the composition of the poor population has shifted. The most dramatic changes have come in the family structure and age distribution of the poor. Between 1960 and the early years of the new century, the proportion of the poor living in female-headed households doubled. Figure 10.4 illustrates the striking reversal that has taken place in the poverty rates of children and seniors, to the advantage of the latter. This shift was, as we will see, the result of federal policies that favored the elderly over families with children. A final, disturbing trend is this: the poor are getting poorer. An easy measure of this is the proportion of people with household incomes under half the poverty level, which has climbed substantially since the late 1970s.

---

[3]Direct comparisons between recent poverty statistics and those for years before 1970 are problematic, for reasons suggested in our discussion of poverty measures. Nevertheless, there is little doubt about the general picture.

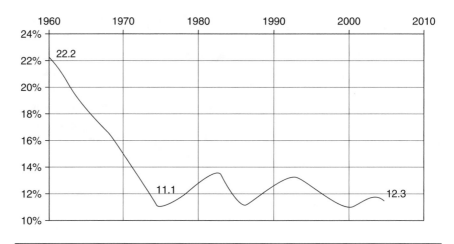

Figure 10.3   Poverty Rates, 1960–2006

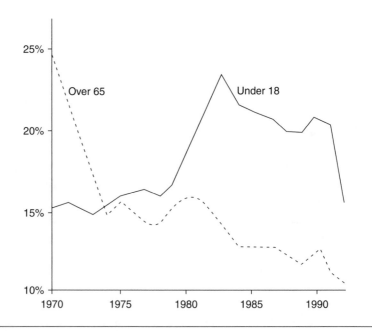

Figure 10.4   Poverty Rates for Children and Seniors

SOURCE: U.S. Census Bureau 2001d.

The Underclass and the Transitory Poor

Knowing how many people are poor from year to year does not tell us how sustained poverty is on an individual level. Many people fall below the poverty line during a given year as a result of temporary circumstances: loss of a job, physical injury, or divorce. When they get back on their feet, they are no longer poor. Others remain poor for many years. They might be disabled, living on meager retirement incomes, or supporting a family with a low-wage job. Some people are just incapable of holding a job for long and thus vulnerable to long-term poverty.

In recent years, interest in sustained poverty has turned into concern that the United States might be developing a permanent underclass—a growing class of people who are impoverished and mired in habits and circumstances that prevent them from ever joining the mainstream. This preoccupation has been fed by the failure of progress against poverty since the 1970s and by the growing concentration of the poor in large cities, where they are most visible to the national media. We will return to the question of the underclass after we examine the few existing studies of turnover in the poor population.

Accurately measuring the duration of individual poverty is difficult. It requires a "panel study"—finding the same individuals to ask the same questions, year after year. The Census Bureau has a continuing survey that tracks people for 2 years. Data from the 1980s and 1990s show that approximately 25% of the people who are poor in a given year are not poor the following year. At this rate, there could be a complete turnover in the poverty population in 4 years. But since the bureau only follows people for 2 years, we have no way of knowing how many people fell back into poverty shortly after climbing out.

The University of Michigan's Panel Study of Income Dynamics (PSID) has been tracking the incomes of a large national sample of families since 1969. The PSID data indicate that brief spells of poverty are common: about a quarter of all families will be poor at least once in a 10-year period, but only a fraction of these families will remain poor for an extended period (Corcoran 1995:241; G. Duncan et al. 1984:41–42).

One PSID study looked at rates of long-term poverty among the families of women age 25 to 44 with children—both female-headed and married-couple families. A woman's family was considered long-term poor if its average income over 6 years was below the poverty line. By this standard, only 5% of such families in the early 1980s were long-term poor (Duncan and Rodgers 1989, as reported in Jencks 1991:35).

These and other studies suggest a rapid turnover in the poor population. Only a minority of poor families are locked into a pattern of long-term poverty. Can we equate this minority, whatever its size, with "the underclass"? That depends on how the term is understood. In the class model we presented at the beginning of this book, underclass was used in a purely economic sense to refer to the poor who are loosely connected to or wholly disconnected from the labor market—encompassing most of the people below the federal poverty line. But in the debate over the underclass, the term is applied to a smaller subset of the poor who are bound to their impoverishment by personal characteristics or structural circumstance. (See Auletta 1982; Jencks and Peterson 1991; Murray 1984; Wilson 1987.)

The underclass label often implies flawed character. The "conventional portrait" of the underclass, according to a *Washington Post* writer, links "extreme poverty, chronic joblessness, welfare dependency, out of wedlock births, female-headed households, [and] high dropout rates" (April 17, 1991). Many would have added

crime to this list. Sociologist William J. Wilson (1987, 1991), one of the most influential writers on the underclass, more or less accepts this grim portrait, but defines
the underclass as having marginal economic position coupled with geographic isolation. Wilson has in mind the inner-city poor, living in areas with extremely high
concentrations of poverty and caught in a postindustrial economic trap of shrinking job opportunities.

Implicit in most discussions of the underclass is the notion of a "cycle of
poverty"—of social pathologies so severe that poverty is inevitably passed from one
generation to the next. Given what we know about social mobility, we would be surprised if there were not at least some truth to this (seldom examined) assumption.
But how much? According to PSID data for the years 1968 to 1988, the vast majority of children raised in poverty are not poor as young adults. As Table 10.3 reveals,
this generalization holds for blacks as well as whites. Only one in four (24.9%) poor
black children grows up to be a poor adult; for whites, the figure is less than 1 in 10.
Childhood poverty does, of course, raise the probability of adult poverty and
depressed adult earnings, as the table indicates. But it is not the grim reaper of life
chances we sometimes assume.

Table 10.3   Intergenerational Transmission of Poverty

	Poverty Rate as Adults (in percent)	Family Income as Adults (in 1990 dollars)
Whites		
Nonpoor as children	1.2	$53,400
Poor as children	9.3	$35,100
Blacks		
Nonpoor as children	9.6	$36,100
Poor as children	24.9	$26,900

SOURCE: Corcoran 1995:247.

NOTE: Refers to adults age 27 to 35 in 1988. Based on stringent poverty measure that counts
food stamps as income and considers average family income level over several years.

Restructuring Welfare[4]

The great entitlement programs enacted or expanded in the 1960s and early 1970s,
such as Aid to Families with Dependent Children (AFDC), food stamps, Medicaid,
and Social Security, helped reduce poverty and improved the lives of the poor. They
also consumed a growing portion of the federal budget and evoked increasing

---

[4]This section draws on Blank 2007; Burtless, Weaver, and Wiener 1997; Congressional
Quarterly 1996; DeParle 1996; Edelman 1997; Grogger and Karoly 2005; Haskins 2006;
O'Neill 2006; and Parrott 2006.

public opposition. Criticism focused on the means-tested "welfare" programs, especially AFDC. Welfare was "growing out of control," critics claimed, and was "full of fraud and abuse."

Presidents from Nixon to Bill Clinton promised to "reform" the welfare system. By and large, they had little luck. Some reform proposals were shelved as unworkable. Others died in Congress. Ronald Reagan, whose political campaigns featured dubious stories of "welfare queens" who had grown rich on public assistance, managed to change the rules to constrain growth in welfare spending, but he left the basic system intact. Clinton courted votes in 1992 with a repeated promise to "end welfare as we know it." His initial, relatively liberal reform plan was transformed by a Republican Congress into a more conservative bill, which Clinton somewhat reluctantly signed in 1996.

The 1996 law, grandly titled the Personal Responsibility and Work Opportunity Act, scaled back three of the major means-tested programs: AFDC, Supplemental Security Income (SSI), and food stamps. The affected programs had long operated as entitlement programs, providing guaranteed assistance to anyone who qualified (under clearly stated rules), backed by open-ended funding. This meant that spending automatically swelled with increasing need in economic bad times and shrank during boom periods. AFDC, the basic public assistance program for needy families with children, had always been a joint federal–state program. SSI was a cash assistance program for low-income elderly, blind, or disabled persons, serving both adults and children. Food stamps provided the broadest safety net protection. The program assisted, in varying amounts, almost anyone whose income did not ensure adequate nutrition, including indigent adults, the working poor, the blind and disabled, low-income seniors, and needy families with children.

Under the 1996 law, AFDC lost its entitlement status and became Temporary Assistance for Needy Families (TANF). *For the first time since the passage of the Social Security Act in 1935, there was no federal guarantee of income assistance for impoverished children and their families.* Now states would receive federal poverty funding in lump sums known as block grants, which they could spend as they wished, subject to two key limitations: (1) Families could not receive more than 5 years of assistance, whether consecutive or nonconsecutive ("Temporary" is clearly the operative term in the program's new title); and (2) most adults benefiting from TANF would be required to begin work of some sort within 2 years of receiving assistance. SSI and food stamps remained entitlements, available to all who qualify. But the law sought to reduce spending on these programs with stiffened eligibility requirements and reduced benefit levels.

The central theme of the legislation was summed up in the oft-repeated phrase "welfare to work." Conservatives who backed the legislation believed that they could end welfare dependency, and perhaps poverty itself, by simply compelling the poor to become self-supporting. There was, moreover, a moral dimension to their expectations. They were convinced that low-income, single mothers who escaped (or avoided) welfare dependency would be more likely to marry and would, if they became self-supporting, provide an example of responsibility that would encourage better behavior in their children. The first, they believed, would improve their chances of escaping poverty, and the second would improve the prospects for the next generation. Many liberals, including Clinton's own welfare experts, were certain that the law would be a disaster for the poor. Some predicted an abrupt upsurge in childhood poverty.

The welfare reform became law at a propitious moment. In 1996, the country was moving into a period of economic expansion and declining unemployment. Shortly before, the cash benefits available to the working poor, especially those with children, under the Earned Income Tax Credit (EITC) were substantially increased—making every dollar they earned more valuable. The minimum wage had just been raised. Other recent legislation enabled poor families to retain health care coverage for their children after they left welfare. (Under the old AFDC program, coverage could be lost—a self-defeating disincentive to work.) Under state programs, child care assistance for working parents was expanded (though still not available to all). In short, in the late 1990s, there were more jobs for the poor and stronger incentives to work.

A decade after the Personal Responsibility and Work Opportunity Act was signed, the results seemed neither as dire nor quite as encouraging as critics and supporters imagined. Researcher Ron Haskins (2006) summed up the most remarkable outcomes of the new system when he told a congressional committee, "The pattern is clear: earnings up, welfare down. This is the very definition of reducing welfare dependency." The national welfare caseload had, in fact, plummeted in the late 1990s, and though it climbed again as the economy slowed in the early 2000s, in 2005, it was still about half what it had been when the act was passed. Over the same decade, though less dramatically, the proportion of single mothers who were employed rose and the childhood poverty rate fell.

These trends confounded the grim expectations of the critics. However, careful research did not reveal the changes in social behavior that conservatives had hoped for, and childhood poverty remained high—around 17%.[5] Among those who left welfare, 60% or less were working, typically for poverty wages of $7 or $8 an hour; especially in the tougher post-2000 labor market, many suffered spells of unemployment. Close observers were especially concerned with the fate of a significant, very poor minority who left TANF—in many cases, forced out of the program because they violated stringent new rules—but had no visible means of support. Many suffered from severe mental or physical problems. A study of long-term TANF recipients (reported in Parrott 2006) who were close to reaching the 60-month lifetime limit for support, found parents who seemed unemployable because of very low cognitive functioning ("could not identify numbers or tell time"), mental illness ("depression so severe that she was unable to maintain basic hygiene"), and physical disabilities ("could not lift a gallon of milk"). Like the minority of AFDC recipients who remain in the system for years rather than just cycling in and out during periods of crisis, those who breach their TANF limits are likely to be people who are least capable of providing for themselves and their families.

---

[5]In 2004, it was 17.8% by the official measure and 16.4% by an alternative measure adjusted for ETIC and other near-cash benefits (Mishel et al. 2007:299).

Clearly, the new welfare legislation and supporting programs like the EITC had, to the surprise of many, reduced welfare dependency without doing significant damage. Even after the economic boom of the late 1990s fizzled, the welfare rolls remained far below where they had been. But the new regime had been less successful against poverty. The official poverty rate remained in the narrow band it had not escaped since the 1970s. Especially disturbing, the proportion of severely poor, with incomes under 50% of the poverty line, was exactly the same in 2005 as it had been the year welfare reform was enacted: 5.4%. In 2005, this figure encompassed 16 million Americans.

The Mystery of Persistent Poverty

Here is the mystery facing students of poverty: Why did the spectacular progress in reducing poverty falter after the early 1970s? Why was the poverty rate in the early years of the new century just about where it had been in the early-1970s, even though the national economy had more than doubled in size over the same period and mean family income was almost 50% higher? There is no simple answer. In this section, we assess three factors: (1) economic change, (2) a societywide revolution in family patterns, and (3) shifts in federal policy.

Economic Trends. In early September 1991, the national news media reported a fire in a North Carolina chicken-processing plant that killed 25 workers and injured more than 50 others. According to the stories, exits that had been locked to keep employees from stealing chicken had contributed to the death toll. In 11 years of operation, state or federal safety officials had never inspected the plant. The workers in the plant were typically working 8 hours a day for $5 an hour or less—a wage insufficient to raise a small family above the poverty line (*New York Times* and *Washington Post*, September 4–11, 1991).

As we noted in earlier chapters, the Age of Growing Inequality has seen a proliferation of such unattractive, low-wage jobs and the loss of many of the better-paying blue-collar jobs. The low skills of those at the bottom of the labor market; long periods of high unemployment; and, more recently, welfare reform have compelled many of the poor and near-poor to accept any available job. One indication of the change was the rising percentage of men earning poverty wages—that is, workers receiving hourly pay so low (under $9.60 per hour in 2005) that even if they managed to work all year full time, they could not earn enough to lift a family of four above the poverty line.[6] The proportion with such low-wage jobs rose from 16% in 1979 to 20% in 2005. The percentage of women earning poverty wages declined substantially during this period but, at 29% in 2005, remained well above that for men (Mishel et al. 2007:125–126).

---

[6]This measure of poverty earnings, based on current hourly wage, is different from the more restrictive standard used for the "U-turn" graph in Figure 3.4, which refers to annual earnings of men who in fact worked all year, full time.

As Figure 10.5 indicates, unemployment rates rose sharply in the late 1970s and early 1980s, and only recently returned to the levels of the 1960s. (The Labor Department statistics on which the graph is based probably underestimate the rise in unemployment during this period.[7]) A glance back at the graph tracing poverty rates (Figure 10.3) reveals that poverty and unemployment rates run parallel courses. In particular, when unemployment spiked during the 1982–1983 recession, the poverty rate reached its highest level in almost two decades. And again, in the milder recession of the early 1990s, both rates rose together. This should not be surprising. Many members of the class we have called the working poor, with annual incomes modestly above official poverty, can easily fall below the poverty line after a spell of unemployment. In short, the labor market turned sour after the early 1970s. Wages went down. Unemployment rose. These developments affected the poor to the extent that they worked or were looking for work.

Just how much do the poor work? Much more, it seems, than we might assume. Of course, many poor people, because they are too young, too old, ill, or disabled, are not expected to work. Almost half of the poor are under 18 or over 65. With these facts in mind, Mishel and his colleagues (2001) focused on the work experience of poor "prime age" adults (men and women from 25 to 54) in 1998, a year when unemployment was relatively low. They found that about 65% worked that year. One in four worked full time, year round. Among poor families headed by a prime-aged adult, 70% of household income came from job earnings. (Only about 10% came from public assistance.) Of course, in many families, more than one person worked in the course of a year.

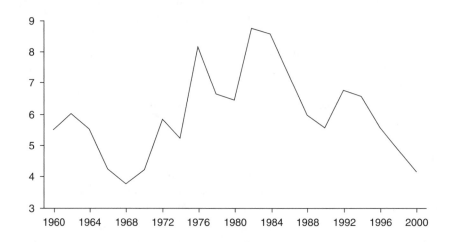

Figure 10.5   Unemployment 1960–2000

SOURCE: U.S. Bureau of Labor Statistics.

---

[7]The department counts as unemployed only people who are not working and have been actively seeking employment. In periods of rising unemployment, some workers become discouraged and stop looking for work, even though they might like to work. In the official count, such people are not part of the labor force and therefore not considered unemployed.

Chapter 10   The Poor, the Underclass, and Public Policy   **221**

The researchers were able to add the total number of hours worked by family members in each poor household (Table 10.4). Among poor families headed by prime-age adults, half worked 1,000 hours or more in 1998. A third of the families worked at least 2,000 hours (the equivalent of fifty 40-hour weeks) without breaching the poverty line. Families with female heads and children put in fewer hours, but even among these households, almost 40% put in more than 1,000 hours.

Table 10.4   Total Hours Worked by Members of Poor Families, 1998

	Percent		
Hours Worked	All Poor Families	With Head 24–54	With Female Head & Children
2000 or more	28	33	16
1000 to 2000	19	21	22
Under 1000	20	19	27
No work	34	27	35
Total	100	100	100

SOURCE: Mishel et al. 2001:320.

Comparisons between 1979 and 1998, using the same data, revealed these sad facts about poverty and work in the Age of Growing Inequality (here we refer to all poor families, not just those with prime-age members): In 1998, more poor families worked. Those that worked put in longer hours for lower real wages, with the net result that they earned fewer real dollars than similar families in 1979.

Although wages remained low for less skilled workers, jobs were generally more plentiful after the mid-1990s. With rising employment, the poverty rate began to decline, demonstrating once again the power of the economy over the poverty level.

Changing Family Patterns. The negative effects of low wages and, for much of the period under consideration, high unemployment have been reinforced by sweeping changes in family patterns. Figure 10.6 summarizes the key trends since 1960. Americans are now less likely to marry and more likely to divorce. Children are six times more likely to be born to unwed parents. Families with children are almost twice as likely to be headed by females.

These trends are pervasive in American society. They affect the poor and the nonpoor, blacks and whites, teenagers, their parents, and grandparents. Together, they amount to a revolution in American family life. The key change in family patterns that sums up the others and has contributed to the rise in the poor population is the increase in the number of female-headed families. Since 1960, the proportion of the poor population living in female-headed families had nearly doubled from 18% in 1960 to 35% in 2005. This development is transforming the lives of children. According to one expert, we have probably reached the point where the majority of children can expect to spend part of their childhood in a

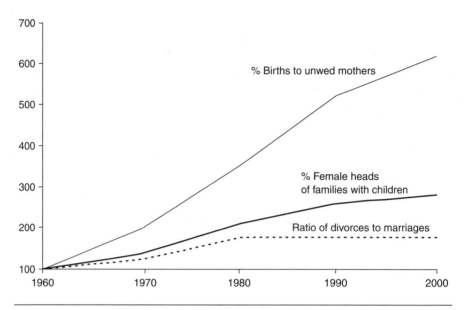

Figure 10.6   Changing Patterns of Family Life

SOURCE: U.S. Census Bureau.

NOTE: All statistics are indexed to a common scale with 1960 equal to 100.

female-headed family, and most of the children who do will experience a period of poverty (Ellwood 1988:45–47, 67).

Why the rapid increase in families headed by women? In part because in the course of the last generation, Americans have become much more tolerant of both unwed motherhood and divorce. Currently, about one-third of births in the United States are to unmarried women. The popular conception of these mothers as irresponsible teenagers is dated and misleading. Girls who become mothers still face tough, poverty-ridden futures; they and their children will be overrepresented among welfare-dependent families. But teen birth rates have declined, probably as the result of greater use of contraceptives and access to abortion. There has been some increase in the number of single-teen mothers because pregnant teens are less likely to marry. But the birth rate for single women in their twenties is higher than that for teens and is growing faster. Women in their twenties account for the majority of unwed births.[8]

One reason for the shift in behavior and attitudes toward divorce, marriage, and single motherhood is that the choices facing women have changed. Job opportunities have expanded, and women are now expected to be capable of supporting themselves. At the same time, the falling wages and erratic unemployment rates of men without special skills or advanced education make them less dependable meal tickets. Sociologists William Wilson and Kathryn Neckerman (1986) present evidence to suggest that these trends have powerfully affected African Americans,

---

[8]Brunner 2001:339; Ellwood 1988:71–72; Jencks 1991:83–93; and current Census Bureau data.

among whom the majority of births are now to unwed women and nearly half of families are headed by women (*New York Times* 2001:285–287). They find that the marriageable pool of employed young black males has contracted sharply, especially since the economic slump of the 1970s. Finally, public assistance became a more dependable source of family support beginning in the 1960s. Although the value of welfare benefits declined after 1970 and was seldom, if ever, sufficient to lift a family above the poverty line, access to public assistance was at least assured for families that qualified until the 1996 reform.

If the array of choices facing young women has expanded, the prospects for single mothers remain unenviable. As we noted in Chapter 4, custodial mothers receive modest child support at best. Employed women, of course, make considerably less on average than employed men. The conflicts between the nurturer and provider roles, which all working mothers face, are even tougher for employed single mothers, who bear the burdens of child rearing alone. Nonetheless, single mothers are more likely to work full time than married mothers. Given the difficulties that female heads of households face, it is scarcely surprising that their families are several times more likely to be poor than dual-headed families (see Figure 10.2) or that the rise in the portion of female-headed families was associated with an era of persistent poverty.

Government Policy. How much can we blame changes in government policy for persistent poverty of the last quarter century? Cuts in various programs serving the poor under President Reagan in the 1980s and the 1996 restructuring of the welfare system under President Clinton certainly made life more difficult for the poor.

But poverty rates had halted their long plunge and begun to move up before Reagan was even elected (see Figure 10.3). Such long-term trends in poverty rates (unlike year-to-year fluctuations) reflect fundamental changes in the national economy and patterns of family life we have been discussing. The government has some power over the economy; it has virtually none over family life.

Some of the federal policies that have most influenced poverty levels are not explicitly poverty policies. For example, the dramatic reduction in poverty among the elderly, visible in Figure 10.4, is the direct result of real increases in Social Security benefits (Levitan 1990:45). The subsequent leveling off of benefits contributed to the reversal of the long-term downward trend in the general poverty rate. Their effect was probably more significant than official poverty rates suggest because of the way poverty rates are calculated (a cut in food stamps, for example, could not affect rates) and because many of the people affected were already poor.

Aside from Social Security, the largest single federal outlay on behalf of the poor is the Earned Income Tax Credit (EITC), which reduces the federal taxes owed by low-income households with job earnings and provides cash payments for working families whose incomes are so low that they owe little or no income tax. With its colorless, forgettable name, the Earned Income Tax Credit functions as a stealth poverty program. It is unknown to most Americans and seldom the subject of public debate. Yet, it is by far the biggest cash transfer program for needy parents and children (Katherin Phillips 2001). Because the EITC is channeled through the tax system, benefits do not show up in poverty rates, which are based on pretax income. (The most sophisticated of the alternative poverty standards we discussed earlier in the chapter does take EITC into account, along with the effects of other tax provisions.)

On the other hand, the failure of the government to uphold the value of the minimum wage and maintain the protections of the unemployment insurance system have certainly pushed some working poor families below the poverty line. In 1970, the minimum wage was high enough that a steady fulltime worker, earning the minimum wage, could keep a family of three above the poverty line; by 2005, the same worker could just earn enough to push one person over the poverty line. Even workers who earn a little more than the minimum wage are affected by it because their wages tend to go up when the minimum wage does. Unemployment insurance can help keep low-income families from falling below the poverty line. But the proportion of the labor force covered, the value of benefits, and the duration of benefits have all fallen significantly since 1975 (Levitan 1990:62–65; U.S. House of Representatives 2000:Section V).

National economic policies also influence poverty rates. Since the early 1980s, national policy has focused on reducing inflation rather than maintaining employment and on increasing profits rather than upholding wages. These goals were reflected in tax and spending policies, as well as monetary policy. International trade policy opened the United States to increasing foreign competition, reducing both employment and wages in the affected industries, especially for workers at the lower end of the labor market. It is difficult to disentangle the results of these policies from the effects of the broad changes in the U.S. economy discussed in Chapter 3.

In sum, persistent poverty since the early 1970s reflects the transformation of the national economy that we have associated with the Age of Growing Inequality and major changes in American family life. The effects of government policy appear to be more modest, though not inconsequential.

Conclusion

This chapter on poverty has, perhaps inevitably, revolved around a series of contentious issues. The first of these concerned the proper definition of poverty. Some writers favor an absolute definition: poverty means not having enough food, proper housing, and so forth. Critics of absolute definitions see them as subjective in practice and largely irrelevant to an affluent society like the United States. These writers prefer a relative definition: poverty means a standard of living far below the mainstream standard of the larger society. The continuing debate about the important technical details of the definition also reflects differences over deeper political questions. As a practical matter, the choice of definition shapes statistical conclusions about the size of the poverty problem and the direction of trends.

The official definition adopted by the federal government in the 1960s and widely used today is an absolute definition. This measure of poverty is based on a material standard much further from the societal average now than when it was created in the early 1960s. As a result, this measure significantly underestimates the poverty in the United States. Nonetheless, government statistics based on the official standard are useful as a guide to differences among segments of the population and to change over time. (See Figures 10.1, 10.2, and 10.3.)

By this measure, some 37 million Americans were poor in 2005. (Alternative standards examined in this chapter ran as high as 54 million.) The official statistics show that the poverty rate declined significantly during the Age of Shared Prosperity. During the Age of Growing Inequality it climbed, but then settled back to about where it had been in the early 1970s. (See Figure 10.3.)

A second issue concerned the idea of a poverty "underclass." We adopted the term as a neutral class label in the model introduced in Chapter 1. But some authors use underclass to refer to a new "dangerous class" supposedly emerging at the bottom of American society. From this perspective, poverty is increasingly a problem of misbehavior: out-of-wedlock births, dependency on government assistance, joblessness, violent crime—all concentrated in high-poverty, often minority neighborhoods. Although the behaviors referred to are certainly part of the larger poverty problem, they do not describe most of the poor. Further, the notion of an inevitable cycle of poverty implicit in discussions of the underclass is not supported by the available data. People who grow up poor are not, by and large, fated to remain poor. In fact, there is a large turnover in the poverty population from year to year. On the other hand, a minority of the poor does seem to be locked into long-term poverty.

The question of the underclass is related to a third issue: the mystery of persistent poverty. Why has progress against poverty faltered since the early 1970s, even though the national economy and the average family income have been growing? Economic trends have certainly played a critical role. For reasons we explored in detail in Chapter 3, the labor market has turned sour for those at the bottom. Even when unemployment rates come down, low-end wages remain quite low. A surprising number of poor Americans are full-time workers. Changing family patterns have also influenced poverty rates. Increasing divorce and out-of-wedlock births, along with the growing economic independence of women, have brought a sharp increase in the number of female-headed families.

As we observed in Chapter 4, most female-headed families are at the lower end of the income distribution, and a large proportion is poor. The truncated educations of many young mothers, the typically lower wages paid to female workers, the frequent failure of absent fathers to pay child support, and the competing demands on single mothers all contribute to this result. But the growth in low-income, female-headed households is, at the same time, a reflection of the changing economic conditions. Many women remain single who might not have done so in the past because the men in their lives are unemployed or working at low-wage, unstable jobs; they do not seem likely to become dependable family breadwinners.

We concluded that the contribution of government policy to persistent poverty is probably marginal relative to long-term economic trends and changes in family life, though some changes in government policies have powerfully affected the poor.

In this chapter, we repeatedly returned to the issues surrounding federal poverty programs. The federal government first took responsibility for the poor during the 1930s under the pressures of the sudden mass poverty brought on by the Depression. A series of protective programs from Social Security to AFDC were created under Franklin Roosevelt's New Deal. In the 1960s and early 1970s, an era of prosperity and protest, poverty returned to the national agenda, most notably under Lyndon Johnson's War on Poverty. Existing programs were expanded and new programs were created—some of them broad insurance programs like Medicare and others means-tested programs like food stamps. Rising Social Security benefits sharply reduced poverty rates among the elderly. The food stamp program reduced hunger and malnutrition.

After Johnson's effort, however, the country abandoned the goal of eliminating poverty. Some politicians substituted a war on welfare for the War on Poverty and found a receptive audience. Means-tested programs were cut back in the 1980s and 1990s. But over the same period, benefits under the Earned Income Tax Credit were substantially expanded, creating a major new source of support for the working poor.

The 1996 law restructuring welfare cut means-tested programs, gave the states wide latitude in their use of federal poverty money, and ended the AFDC entitlement. For the first time since 1935, there is no guarantee of income support for impoverished children and their families. The act yielded a stunning reduction in the welfare rolls, but many of the people who left the system or were pushed out of it by time limits or stringent new rules were desperately poor and unable to provide for themselves, even in the best of times.

Key Terms Defined in the Glossary

absolute poverty standard
  (see poverty standards)
Aid to Families with
  Dependent Children (AFDC)
Earned Income Tax Credit (EITC)
entitlement
government transfers
in-kind benefits
means-tested programs
minimum wage
125% of the poverty line
poverty line (see poverty threshold)

poverty rate
poverty standard, federal (official)
poverty standards, absolute and relative
poverty threshold
relative poverty standards
  (see poverty standards)
risk of poverty
Temporary Aid to Needy Families
  (TANF)
underclass
working poor

Suggested Readings

Anderson, Elijah. 1992. *Streetwise: Race, Class, and Change in an Urban Community*. Chicago: University of Chicago Press.
    Two highly regarded ethnographic studies of the urban underclass.

————. 1999. *Code of the Street: Decency, Violence, and the Moral Life of the Inner City*. New York: Norton.

DeParle, Jason. 2004. *American Dream: Three Women, Ten Kids and a Nation's Drive to End Welfare*. New York: Penguin.
    Three cousins, their families, and welfare reform. DeParle, a New York Times *reporter, begins the family history on a Mississippi plantation before the Civil War.*

Edin, Kathryn and Maria Kefalas. 2005. *Promises I Can Keep: Why Poor Women Put Motherhood Before Marriage.* Berkeley: University of California Press.
> *A 5-year study of 162 young, poor single mothers.*

Edin, Kathryn and Laura Lein. 1997. *Makin' Ends Meet. How Single Mothers Survive: Welfare and Low-Wage Work.* New York: Russell Sage Foundation.
> *How low-income, single mothers make and spend money. Based on extensive interviews and detailed budgets. An important, troubling book that places welfare "reform" in a practical context.*

Fine, Michelle and Lois Weis. 1998. *The Unknown City: The Lives of Poor and Working-Class Young Adults.* Boston: Beacon.
> *Family and economic lives of white, black, and Hispanic young adults in the context of two economically troubled cities in the Northeast, most working poor or underclass.*

Piven, Frances Fox and Richard A. Cloward. 1971. *Regulating the Poor: The Functions of Public Welfare.* New York: Pantheon.
> *Stimulating analysis of the development of welfare programs in the context of the social and political forces that spawned them.*

Shipler, David. 2005. The Working Poor: *Invisible in America.* New York: Random House.
> *The lives of the working poor. A sprawling, insightful book by a former* New York Times *reporter.*

Snow, David and Leon Anderson. 1993. *Down on Their Luck: Homeless Street People.* Berkeley: University of California Press.
> *Ethnography of the daily lives of homeless people in Austin, Texas.*

Venkatesh, Sudhir Alladi. 2000. *American Project: The Rise and Fall of a Modern Ghetto.* Cambridge, MA: Harvard University Press.
> *Study of a Chicago inner-city housing project—the forces that shaped it and the lives of its residents.*

Wilson, William Julius. 1987. *The Truly Disadvantaged: The Inner City, the Underclass, and Public Policy.* Chicago: University of Chicago Press.

———. 1996. *When Work Disappears: The World of the New Urban Poor.* New York: Knopf.
> *Two influential books on the origins and nature of ghetto poverty.*

CHAPTER 11

The American Class Structure and Growing Inequality

[We have seen] the triumph of upper America—an ostentatious celebration of wealth, the political ascendancy of the richest third of the population and the glorification of capitalism, free markets and finance. But while money, greed and luxury [became] the stuff of popular culture, hardly anyone asked why such great wealth had concentrated at the top.

Kevin Phillips (1990)

I n this final chapter, we synthesize what we have learned about the American class system and how it is changing. We review the evidence we have found of increasing inequality and reconsider the possible reasons for this trend. Readers may notice a shift of tone. In the preceding chapters, our overriding objective was to present the existing data and research as precisely and faithfully as possible. Here we are less constrained. We generalize broadly, emphasizing our own interpretations. We, by and large, dispense with citations, numbers, and tables—to the relief, no doubt, of many readers.

How Many Classes Are There?

Those with good memories will recall our answer to this question from the first chapter: six, but it all depends on your viewpoint. There is no irrefutable answer. As we noted in Chapter 1, defining classes and specifying the dividing lines between them is as much art as science. The "teardrop" class model we developed in Chapter 1 (see Figure 1.1, p. 13) reflects what we have learned researching this book, but it inevitably imposes a simplified pattern on a complicated, even chaotic reality. Let's take a closer look at the model and how it was derived.

Drawing on Marx and Weber, we began with the assumption that the class structure develops out of the economic system. Our classes are based on economic distinctions, rather than the prestige distinctions that Warner and his successors employed to develop their class models. But because prestige is generally rooted in economic differences, our map of the class system is broadly similar to Warner's prestige model for Yankee City or Coleman and Rainwater's model for Boston and Kansas City. There are two major differences: We do not employ their new money/old money distinction at the top, nor do we treat blue collar/white collar as the essential class distinction in the middle (Table 2.2, p. 30, compares the three models).

As we learned in Chapter 4, three basic sources of income are available to households in this country: capitalist property, job earnings, and government transfers. The first source allows us to distinguish a top class that largely depends on income from property. The last source points to a class that has virtually no property and limited labor force participation, but often depends on government transfers. Those who fall between these class extremes rely on their jobs—ranked by the skill or education required, the judgment and authority exercised at work, and the level of earnings.

Taken together, these factors suggest a structure of six classes:

1. A *capitalist class,* subdivided into nationals and locals, whose income is derived largely from return on assets.

2. An *upper–middle class* of college-trained professionals and managers (a few of whom ascend to such heights of bureaucratic dominance or accumulated wealth that they become part of the capitalist class).

3. A *middle class* whose members have significant skills and perform varied tasks at work, under loose supervision. They earn enough to afford a comfortable, mainstream lifestyle. Most wear white collars, but some wear blue.

4. A *working class* of people who are less skilled than members of the middle class and work at highly routinized, closely supervised manual and clerical jobs. Their work provides them with a relatively stable income sufficient to maintain a living standard just below the mainstream.

5. A *working-poor class,* consisting of people employed in low-skill jobs, often at marginal firms. The members of this class are typically laborers, service workers, or low-paid operatives. Their incomes leave them well below mainstream living standards. Moreover, they cannot depend on steady employment.

6. An *underclass,* whose members have limited or erratic participation in the labor force and do not have wealth to fall back on. Many depend on government transfers.

The cutting points suggested by this schema are not equally significant or salient. The capitalist class is clearly separated from other classes by a crucial distinction: wealth as a primary source of income. The upper-middle class is set off from the classes below it by valuable credentials and the rewards that flow to them. The underclass is isolated from the classes above it by its loose connections with the world of work. But the other class boundaries are not so neatly defined.

The hazy distinction between the middle and working classes, the two great classes at the center of the class structure, has long perplexed students of stratification. We have ignored the traditional blue-collar/white-collar distinction and emphasized differences of education, skill, and autonomy connected with particular occupations. Thus, the electrician is middle class and the clerical worker doing routinized office tasks is working class, despite the colors of their shirts.

In summary, we are suggesting a model of the class structure based primarily on a series of qualitative economic distinctions. Thus, we emphasize the source of income and pay less attention to the amount of income. Our schema is summarized in Table 11.1, which provides fuller detail than the teardrop graphic in Chapter 1. As the table indicates, we think of the six classes as divided into three broader categories: the privileged classes (capitalist and upper middle), the majority classes (middle and working), and the lower classes (working poor and poor). The percentages in the table are rough estimates of the proportion of households in each class, based on the available occupation, income, and wealth data. (Our rationale for conceiving of classes as groupings of households was presented in Chapter 1.) The classes are defined by the sources of income and occupations listed in the third column. Occupation here refers to the work of the highest-earning member of the household. The last two columns list education levels and incomes typical of each class. Of course, both will vary considerably in practice. Household income, as we learned in Chapter 4, is very dependent on the number of family members in the labor force. Educational levels have risen in successive generations, so that a younger worker will tend to have more years of school than an older worker in the same class.

Let's take a closer look at the six classes.

Table 11.1   Model of the American Class Structure: Classes by Typical Situations

Class, Percent of Households	Source of Income, Occupation of Main Earner	Typical Education	Typical Household Income, 2005
Privileged Classes			
Capitalist 1%	Investors, heirs, executives	Selective college or university. Often post-graduate	$2 million
Upper Middle 14%	Upper managers and professionals, medium-sized business owners	College, often post-graduate study	$150,000
Majority Classes			
Middle 30%	Lower managers, semi-professional, nonretail sales workers, craftsmen	At least high school, often some college	$70,000
Working 30%	Operatives, low-paid craftsmen, clerical workers, retail sales workers	High school	$40,000
Lower Classes			
Working Poor 13%	Most service workers, laborers, low-paid operatives, and clerical workers	At least some high school	$25,000
Underclass 12%	Unemployed or part-time workers, many dependent on public assistance and other government transfers	Some high school	$15,000

The Capitalist Class

The members of the tiny capitalist class at the top of the hierarchy have an influence on economy and society far beyond their numbers. They make investment decisions that open or close employment opportunities for thousands of others. They contribute money to political parties, and they own media enterprises that allow them influence over the thinking of other classes.

The capitalist class strives to perpetuate itself: Assets, lifestyles, values, and social networks are all passed from one generation to the next. (In Bourdieu's terms, economic, cultural, and social capital are all vital parts of the inheritance.) The maintenance of family fortunes and cohesion becomes more difficult as kin multiply and holdings are divided. Families attempt to instill a sense of lineage in the young. Extended families are drawn together by regular contact and by mutual dependence on their shared estate. Members of this class are active supporters of preparatory schools and colleges for their children and for ambitious newcomers whom they hope to see socialized into their worldview.

The richest, most powerful members of the capitalist class operate on the national and international scene. Some control major corporations. They donate large sums to political campaigns and other political projects. They fund foundations and public policy "think tanks." These national capitalists have less prominent counterparts in communities across the country—the people who own the local car dealerships, real estate empires, media outlets, and other major local businesses. They fund community nonprofit institutions and influence local politics. Their collective influence over national politics is considerable because they are likely to have easy access to their own members of Congress (whose campaigns they help finance) and belong to politically potent local organizations like the Chamber of Commerce. At both the national and local levels, the political power of the capitalist class has grown since the late 1970s.

The capitalist class is defined by dependence on income-producing wealth. We include owners of substantial enterprises, investors with diversified wealth, heirs to family fortunes, and top executives of major corporations. Top executives are included because their multimillion-dollar compensation typically includes a stake in the company they manage and permits the accumulation of sizable personal fortunes.

The organization of wealth in this country has been changing for decades. Family-controlled enterprises still account for a large share of capitalist class wealth and income. Among the largest corporations, however, family control has, by and large, given way to a system of control by professional executives. As this happens, members of the capitalist class are diversifying their holdings. At the national level, families are less likely to be identified with a single enterprise. Locally, families are selling the enterprises with which they established their fortunes to national corporations. Local banks and retail stores, newspapers, and television stations are being absorbed or displaced by national firms. The younger generation inherits diversified stock and bond portfolios. Local wealth becomes national wealth.

These changes create the basis for a more cohesive capitalist class, whose members are free from parochial identification with a particular firm, economic sector, or locality. In politics, this tendency has been reinforced since the 1970s by the business PACs, business lobby groups, and policy research organizations that defend the interests of the capitalist class as a whole rather than those of individual capitalists.

The Upper–Middle Class

Since the early decades of the twentieth century, the upper–middle class has grown in numbers and importance and its composition has changed. Increasingly, salaried managers and professionals have replaced individual business owners and independent professionals. The key to the success of the upper–middle class is the growing importance of educational certification in a society dominated by complex technology and large-scale organization. Weber, who died in 1920, saw this coming. He observed the spread of the bureaucratic form of organization, with its characteristic preference for formal credentials, in business, government, and elsewhere. Weber (1946) compared the "preferential social opportunities" available to the university-educated in modern societies to the privileges of the well born in aristocratic societies (pp. 241, 301).

At the very top of the upper–middle class, we find a small but growing stratum that we call the working rich. Its typical members are very successful professionals (doctors, lawyers, dentists) and ranking (but not top) corporate executives—all with incomes in the hundreds of thousands of dollars. We originally encountered them near the end of the income parade. Though rich, they cannot be considered members of the capitalist class because their incomes are largely generated by professional fees or executive salaries, rather than income-producing assets. Like the working poor, they have jobs and depend on them.

The upper–middle class exercises enormous and growing influence in American society. Because the upper–middle class embodies American achievement ideas and possesses vast purchasing power, its lifestyles and opinions are becoming increasingly normative for the whole society. The members of this class vote, volunteer, make campaign donations, and run for office. Their high level of political activity amplifies the influence of their political views and electoral choices.

We have described the upper–middle class as one of the two "privileged classes." It is, in fact, a fairly porous class, open to people of modest origins who manage to earn the right credentials. But at the same time, it is increasingly separated from the rest of society. The income gap between the upper–middle class and the nonprivileged classes has widened. Upper–middle class families are more likely to live in class-segregated neighborhoods and send their kids to class-segregated schools. We are convinced that the gap between this class and the rest of the population has replaced the traditional blue-collar/white-collar division as the most important cleavage in the American class structure.

The Middle Class

The stratification hierarchy, as we have repeatedly observed, is clearest at the extremes. Toward the center, distinctions become blurred, people move more often during their lives from one level to another, and status becomes ambiguous. This is particularly true at the point where the middle class and the working class intersect—or better, overlap—so the reader should not expect precision of classification.

The changing character of work has largely eliminated the traditional differences between blue-collar and white-collar employment. The declining income differential between them, the increasing routinization of clerical tasks, and the corresponding drop in the prestige value of a white collar per se have all helped close the gap between shop and office.

Viewed in terms of major occupational groupings, the problem of distinguishing the two majority classes centers on the sales, clerical, and craft categories. We distinguish jobs that require little preparation, are highly routinized, closely supervised, and typically not well paid from jobs that demand significant skill or knowledge, are fairly varied, autonomous, and better rewarded. On this basis, semiprofessionals (teachers or social workers, for example), low-level managers, most craft workers, and foremen are all middle class. Operatives, such as semiskilled factory workers and truck drivers, belong in the working class, along with routine clerical workers, whose jobs often have an assembly-line character and are not well rewarded.

Government statistics divides sales occupations into retail and nonretail jobs, which we categorize as working and middle class, respectively. The nonretail group includes insurance salespeople, real estate agents, and manufacturers' representatives. A few highly paid sales workers—stockbrokers, in particular—should be considered upper-middle class. On the other hand, we place the lowest-paid sales, clerical, and operative positions in the working-poor class.

By most definitions, the middle class is large and diverse. The first characteristic makes it a natural target of political campaigns; it is difficult, in most districts, to win elections without middle-class support. But the diversity of the middle class makes for ambiguous, if not conflicting, political interests. Although its numbers are larger, the middle class probably has less political influence than the upper-middle class, which is much more active politically and has a clearer sense of political direction.

"The disappearing middle class" has become a recurrent theme for political observers, journalists, and pop sociologists. In fact, the middle class is probably growing and, given the occupational trends associated with postindustrial society, is likely to grow in the future. Of course, the middle class can be made to disappear and reappear by manipulating the way it is defined. From our perspective, it is not the middle class but rather the middle-income group that is shrinking because of declining earnings of many working-class positions and the growth of family incomes toward the upper end of the distribution.

The Working Class

The traditional core of the working class is easy to identify: semiskilled machine operatives, in factories and elsewhere, whose proportion of the labor force has been shrinking since the 1950s. Workers in routine white-collar jobs are, by our definition, also part of this class and their share has been growing. Working-class households often combine blue-collar, white-collar, and service jobs.

The loss of many goods-producing jobs has powerfully affected the working class and the country as a whole. New high school graduates can no longer be assured of finding jobs that will enable them to support families. The contraction of employment in goods-producing industries has helped reduce union membership to a small fraction of the labor force. This trend has, in turn, reduced the political influence of the working class and reinforced the conservative trend in national politics.

Members of the working class have low rates of political participation. With the decline of the urban political machines and labor unions that once mobilized them, they are less likely to vote than they were as recently as the 1960s.

The Working Poor

The working-poor class includes most service workers[1] and the lowest-paid operatives and sales and clerical workers. Aside from low pay, their jobs generally have other disadvantages, such as unpleasant or dangerous work, few benefits, and uncertain employment. The distinctions between this class and the classes below and above it are both problematic. Many jobs near the boundary between the working class and the working poor could be placed on either side. But the gap between average jobs in the two classes is quite large. Lifestyles also differ. The working poor tend to have less stable work histories—sometimes for reasons beyond their control—and more personal and family problems. The lower boundary of this class, defined by commitment to employment, is blurred by the tendency of some individuals to move back and forth across it—a pattern that might be described as oscillating mobility.

Many of the working poor are young workers who, with further training and experience, will be able to move up in the class hierarchy. But income studies that follow families for extended periods reveal that a growing proportion of low-income families are experiencing stagnating or declining fortunes. Like the underclass beneath them, the working poor are generally alienated from political life and have minimal influence on the political process.

The Underclass

Low-income households whose members have limited participation in the labor force form the underclass. They may work erratically or at part-time jobs, but their lack of skills, incomplete education, spotty employment records, and—in many cases—disabilities make it difficult for them to find regular, full-time positions. Many are single mothers, receiving uncertain support from absent fathers, balancing

---

[1]Among the exceptions are police and firefighters, classified as service workers by the Census Bureau, but as middle class by our criteria.

nurturer and provider roles, and facing a job market that offers little to women with limited education. A significant proportion of the underclass depends on government transfer programs, including public assistance, Social Security, Supplemental Security Income (SSI), and veteran's benefits. The 1996 law restructuring welfare is making life more difficult for some of the most vulnerable members of this class, though its long-term results are still difficult to gauge.

Growing Inequality

A key theme in this book, announced in the subtitle, has been the expansion of class disparities in recent decades. The U-shaped curves we saw in Chapter 1 trace a momentous shift, somewhere in the early 1970s, from increasing equality to rising inequality. On this basis, we distinguished two periods: the Age of Shared Prosperity (1946 to the early 1970s) and the Age of Growing Inequality (the period since the early 1970s). The date dividing these periods is imprecise because the exact timing depends on the indicator chosen, but the general pattern is unmistakable in the data on wealth, income, earnings, poverty, and other measures we have examined (see Figure 1.2, pp. 15–16) for a graphic overview.

In the remaining pages of the book, we will summarize the trends we found and return to the question we have asked repeatedly: Why is this happening?

Occupational Structure. Around 1970, the United States completed the transition from an industrial to a postindustrial society—that is, from a society in which most workers were employed in predominantly goods-producing sectors of the economy to one in which most were employed in service-producing sectors (see Table 3.4, p. 56). We found that the postindustrial economy tends toward occupational polarization. It requires engineers, accountants, and physicians, but it also needs janitors, cashiers, and hospital orderlies. The demand for operatives, miners, and other direct goods producers shrinks in the postindustrial economy, thereby eliminating some of the better-paying positions available to workers without sophisticated educational credentials.

Earnings. Job earnings provide one of the best measures of the trends we have been discussing. During the Age of Shared Prosperity, rapid economic growth produced a steady rise in the average wage. But since the early 1970s, wages have more or less stagnated and the distribution of wages has become increasingly unequal. The change has been most dramatic at the top and bottom: between 1980 and 2000, the real compensation of CEOs grew 600%, while wages at the bottom of the labor market were sinking.

The trend toward earnings inequality is remarkably pervasive. Inequality has increased between the college and high school educated, between the skilled and the unskilled, between younger workers and older workers, but also among the college educated, among the high school educated, among doctors, among carpenters, among people employed in manufacturing, among those employed in the service sectors, and so on. Even among corporate executives, inequality has increased, as

CEO earnings surged ahead of the rewards available to vice presidents and middle managers. There appears to be only one notable exception to this pattern: the earnings gap between men and women has narrowed.

Wealth. Inequality of wealth, measured by the concentration of net worth in the top 1% of the population, declined during the 1960s, only to rise sharply after the mid-1970s. In a sense, wealth represents a deeper inequality than earnings or income: it can be passed from generation to generation; it can provide such life fundamentals as a college education or home ownership; and, in capitalist form, it can generate new income. The increased concentration of wealth suggests mounting inequalities in the future.

Poverty. The poverty rate, as officially measured, plunged during the 1960s, but more or less stagnated thereafter. In 2005, it was about where it had been in the late 1960s, although the size of the national economy had more than doubled and mean household income had grown by a third. The officially poor were further than ever from the mainstream. In the last quarter century, the makeup of the poverty population changed. The poor are less likely to be over 65 and are more likely to be under 18 than they were in 1960. They are also much more likely to live in female-headed families.

Income. Income provides the broadest and most continuous measure we have of economic inequality. Family income trends unambiguously reveal the differences between the Age of Shared Prosperity and the Age of Growing Inequality. During the first period, incomes at all levels rose swiftly and the distribution gradually moved toward greater equality. Remarkably, from 1950 to 1975, the real incomes of the poorest 40% of families almost doubled. In contrast, from 1975 to 2000, incomes grew very slowly at the bottom and in the middle, but soared at the top. In fact, the average income of the top 5% of families more than doubled. (See Figure 4.7, p. 88.)

Social Life. Wealthy Americans have always been able to insulate themselves from the grimy realities of life at lower levels of the class structure. Has the increased economic inequality of recent decades meant increased social inequality? One important piece of evidence suggests that it has. The decennial census shows increased residential segregation by income level from 1970 to 2000. In particular, high-income families are increasingly likely to live in separate neighborhoods. (In contrast, earlier studies indicate that residential segregation by class was decreasing in the 1960s.) This residential separation means more schools, malls, and soccer leagues that are class-segregated, and fewer settings like the fictional neighborhood bar in *Cheers,* where people of different classes encounter each other as equals.

Social Mobility. Intergenerational social mobility slowed in the Age of Growing Inequality. There was less upward and more downward mobility at the beginning of a new century than there had been a generation earlier. The decline was most evident among younger workers. Nonetheless, even among younger workers, upward mobility was still more common than downward mobility. And even at the extremes of the class structure, position is not fixed at birth. For example, most young adults who were poor at birth are no longer poor (though their incomes typically remain below average), and a large percentage of the superrich Americans on the *Forbes* 400 list were not wealthy as children (though their families may have been well-off).

In short, in the Age of Growing Inequality, it's harder to get ahead, but upward mobility is still pretty common. Does slowing mobility produce increasing inequality? Not necessarily. For example, if everyone moved down a notch (or up a notch), all would remain in the same relative position—inequality would not be affected. Recent studies do not suggest that a changing pattern of intergenerational mobility is contributing to increased inequality.

Political Power. In the Age of Growing Inequality, power has shifted away from the working class and working poor and toward the privileged classes—in particular, the capitalist class. Of course, power cannot be measured like wealth or income, and classes are not conscious political actors—they are abstractions. Yet, we conclude that the class balance of power has shifted because the relative influence of institutions representing the interests of different social classes has changed and because national decision making has become more favorable to the privileged.

In the early 1970s, national business leaders initiated a concerted drive to expand their influence in national affairs. With money from corporations and wealthy individuals, they organized to protect business interests, promote conservative ideas, and back conservative candidates. Frequently, these candidates represented upper–middle-class constituencies, thus forging an implicit political link between the two privileged classes. During the same period, labor unions were in a steep decline, losing members, money, and political power. The unions had long spoken for a wider working-class constituency than their own members. The business political initiative and the decline of organized labor were among the key factors that contributed to a series of critical victories for conservatives beginning with the election of Ronald Reagan in 1980. Political developments from reductions in the tax rates on very high incomes to the defeat of efforts to create a national health care system reflect the shifting class balance of power.

Why?

How, then, can we account for the Age of Growing Inequality? No one knows exactly, but here's a short, provisional answer: There were some big changes in the economy. The effects of these economic changes were amplified by the decisions of corporations, families, and government.

The onset of the Age of Growing Inequality roughly coincided with the transition from an industrial to a postindustrial society. As we have seen, wage disparities of all sorts have widened. One reason is that disparities are greater in the growing service-producing sectors of the economy, such as restaurants, health care, and law, than in the shrinking goods-producing sectors, such as manufacturing, mining, and transportation. The new economy (in both goods-producing and service-producing sectors) makes winners out of workers with advanced education and skills, and losers out of those who lack such training. In part, this is because of the effects of technological change and globalization. Advancing technology increases the demand for engineers, scientists, technicians, and those who manage technology, while reducing demand for crafts workers, operatives, and laborers. The ease with which capital and goods now move around the globe favors the investor, the accountant, the aeronautical engineer, and the systems analyst, but undercuts the factory worker. Shoes, textiles, clothing, and consumer electronics products can be made by low-wage labor in Mexico or China.

Institutional mechanisms that once constrained wage differentials have weakened since the early 1970s. For example, fewer workers are protected by labor unions and collective-bargaining agreements. The value of the minimum wage has been allowed to erode. Federal deregulation seems to have resulted in increased pay differentials in industries such as airlines and trucking.

Facing more competitive markets at home and abroad, corporations look for ways to cut labor costs. In the Age of Growing Inequality, corporations "downsize," "outsource," and move production abroad. They eliminate benefits, freeze wages, and create new classes of workers, including part-timers, temporary workers, second-tier new hires, and leased workers. They develop union-avoidance strategies. Although managers and professionals have not escaped the layoffs and wage reductions, these "meaner, leaner" labor practices have had their biggest impact at lower occupational levels. Corporations still offer generous competitive rewards to those whose talents are important to their success, from software designers to store managers—and, of course, CEOs.

Wage earners are family members, and the effects of economic change are refracted through the prism of changing family life. In the new era, Americans are less likely to marry, more likely to divorce, more likely to have children out of wedlock, and, as a result, much more likely to live in female-headed families. Because absent fathers typically pay limited child support and women generally earn less than men, female-headed families are concentrated on the low end of the income distribution. Among lower-class men and women, the failure to marry may itself reflect the reduced capacity of young men with limited education to provide for families as they might have in the past.

In the Age of Growing Inequality, more women are employed, they work longer hours, and they earn higher wages. Women's rising earnings have enabled families to overcome the effects of erosion in men's earnings. But this trend is also contributing to disparities in family incomes. Families depending on one worker—male or female—typically fall well behind two-paycheck families. High-earning women tend to be married to high-earning men and to experience faster growth in earnings. The net result of women's increased employment is greater inequality among families.

Relative to the giant consequences of change in the economy and family life, the effects of government policy have probably been modest, but not insignificant. We have already referred to the failure of the federal government to maintain the value of the minimum wage and to the effects of deregulation. The decline of the labor movement is, in part, the result of weakened federal protection of union rights and of antiunion laws in states that provide a refuge for firms searching for cheap labor. Recent legislation has weakened the social safety net protecting the poor. Presidents from both parties have promoted the elimination of barriers to international trade, a policy that has probably been injurious to the lowest-skilled workers, whatever its advantages to the economy as a whole.

All these policies allow greater freedom for market forces, which tend to produce unequal outcomes: rising rewards for some who are talented, well financed, or just lucky, and stagnant or declining rewards for many others. In broad terms, public policy in the Age of Growing Inequality has become more responsive to the privileged classes and less sympathetic to other classes. Tax policy, for example, has zigged and zagged since the early 1970s, but the most significant net result has been sharply reduced federal taxation of wealthy households. Macroeconomic policies reflect the same tendency.

Federal policy makers have been more concerned with reducing inflation than with maintaining employment or wage levels. In the 1990s, as unemployment declined, senior federal officials strained to reassure worried investors that wages were not rising as a consequence. No one bothered to reassure low-income workers about depressed wages. Government might have cushioned the effects of economic change for those most adversely affected. It might have created new opportunities for people who lacked the education or skills to compete successfully in the new economy. But it has, by and large, failed to do so.

Vital Signs

In the first decade of the new century, the American economy was growing fairly steadily, unemployment was generally low, and labor productivity was rising. Such conditions should produce higher wages, but the median wage was stagnant. Only at the top of the labor market were wages rising. Family incomes followed the same pattern. The benefits of growth flowed to the top 20% of workers and the top 20% of households. After pausing briefly at the beginning of the decade, CEO compensation resumed its multimillion-dollar ascent. But economic growth had minimal effect on the poverty rate, which was higher in 2006 than it had been in 2000—higher, in fact, than it had been in 1976. In Washington, the estate tax was placed on the path to extinction, and the minimum wage, eroded by inflation to near irrelevance, was finally raised, but to a level lower in real terms than the minimum wages of the 1960s and 1970s. There was, after three decades, little to suggest that the end of the Age of Growing Inequality was near.

Glossary

Note: Terms in bold within definitions are listed separately in the glossary.

Absolute Poverty Standard   See **poverty standards.**

Age of Growing Inequality   The period beginning in the mid-1970s when inequality rose on multiple dimensions including wages, income, and wealth. The growing inequality of this period is contrasted with the economic growth of the preceding **Age of Shared Prosperity.**

Age of Shared Prosperity   The years from the end of World War II to the early 1970s, when incomes at all levels grew at a healthy pace and economic and social inequalities were declining. Contrasted with the subsequent **Age of Growing Inequality.**

Agricultural Society   See **postindustrial society.**

Aid to Families with Dependent Children (AFDC)   The **means-tested** cash assistance program designed to aid needy, typically female-headed, families with children. Replaced in 1996 by **Temporary Aid to Needy Families (TANF).**

Alienation   In Marx, a sense of powerlessness experienced when human beings lose control of their own creations and even become subject to them. One aspect of this is alienation at work—for example, the assembly-line factory worker who must create something that someone else designed, that someone else will use, and at a speed determined by someone or something else.

Association   Patterns of interpersonal contact, such as in leisure activities, friendship, and marriage, especially among members of the same class. Class subcultures, marked by common values and lifestyles, can emerge when people of similar class position associate more often with one another than with persons of lower or higher classes. See Chapter 5.

Blue-Collar Workers   Manual workers including crafts workers, **operatives,** and laborers. Sometimes used as shorthand for the **working class.** Distinguished from **white-collar workers** and **service workers.**

Bourgeoisie   Marx's term for the class that owns the means of production and controls the **superstructure** in a capitalist society.

Capital   The funds, goods, machinery, land, and so on, invested in an enterprise by its owners. Bourdieu (see Chapter 5) extended this traditional conception to encompass three forms of capital: *economic capital,* the basic monetary form, institutionalized as property rights; *cultural capital,* knowledge in its broadest sense, institutionalized as educational credentials, but encompassing such matters as table manners and how to swing a tennis racket; and *social capital,* mutual obligations embodied in social networks such as kinship, friendship, and group membership.

Capitalism   An economic system based on private ownership of business (capital) and controlled by markets in which **capital,** labor, goods, and services are freely bought and sold.

Capitalist Class   In the **Gilbert-Kahl model,** the very small top class composed of people whose income is largely derived from return on assets. In Marx, same as **bourgeoisie.** See Chapter 8.

Chain of Causation   A series of successive causal influences, especially in occupational achievement. For example, father's education may influence son's education, which in turn influences son's occupation.

Chief Executive Officer (CEO)   Highest ranking executive in a corporation.

Circulation Mobility   Mobility made possible by movement within the existing occupational structure. For example, if some offspring of men with high-status jobs take lower-status jobs, they create opportunities for the offspring of lower-status jobholders to move up. See **structural mobility** and Chapter 6.

Class Consciousness   The recognition by the members of a class of their common identity and shared interests. Marx saw class consciousness as a precursor to class conflict and revolution. Modern social scientists are more interested in class consciousness as an influence on political opinion, electoral preferences, and labor militancy. See Chapter 9.

Class Identification   The class label people choose for themselves, particularly in response to some variant of the standard survey question: Do you consider yourself upper class, middle class, working class, or lower class? Responses consistent with **objective class position** are indicative of an important aspect of **class consciousness.** See Chapter 6.

Class Perspective   See **elite; pluralist perspective**; and Chapter 8.

Class Position, Objective   Position in the class structure as determined by objective criteria selected by the analyst. Contrasted with the subjective consciousness individuals have of their own class position. For example, an observer may define all who work for wages at manual jobs as working class (objective class position), but those who hold such jobs may include some who regard themselves as middle class and others who see themselves as working class (subjective consciousness).

Correlation, Simple   A coefficient indicating how accurately the value of one variable can be predicted from another. A coefficient of 0.0 indicates that there is no relationship between the variables. A coefficient of +1.0 or −1.0 indicates that the value of one variable can be perfectly predicted by knowing the value of the other, because the two values move up and down (or in opposite directions) in unison. In the social sciences, most correlations are intermediate. For example, the correlation between father's education and son's education is 0.45, suggesting very significant, but not consistently decisive influence of the first on the second.

Cultural Capital   See **Capital.**

Cutting Consistency   The degree of community consensus about the points at which a hierarchy of individuals, families, or occupations should be divided into social classes. Studies have shown there is greater consensus about rankings than division into classes. See **ranking consistency** and Chapter 2.

Distribution of Income   The pattern of income inequality, typically presented in one of two formats: (1) the distribution of households across income levels (how many households earn between $0 and $10,000, $10,001 to $15,000, etc.) or (2) the division of shares of aggregate income among fractions of the population (what percentage of all income is received by the richest fifth of households, the second fifth, and so on). See Chapter 4.

Distribution of Wealth   The pattern of wealth inequality, typically measured as the proportion of all **wealth** (or net worth) owned by some percentage of the population. See Chapter 4.

Dividend   A portion of a company's profits periodically paid to its shareholders in proportion to the number of shares each owns.

Downsizing   Large-scale layoffs by corporations to lower costs and boost profits. Notable in the United States since the 1980s.

Earned Income Tax Credit (EITC)   A provision of the federal tax code reducing the taxes for low-income families with job earnings. Those who qualify, but owe little or no federal income tax, may receive the credit as a cash payment. An important supplement to the incomes of many poor families.

Earnings   Money received as payment for doing a job or as profit from a small business or professional practice. One of the components of **income.**

Economic Capital.   See **Capital.**

Effective Tax Rates   The proportion of household income lost to the combined effect of all taxes (income tax, payroll taxes, estate tax, and so on). Typically used in connection with federal taxes. Since taxes vary considerably in their incidence at different income levels, effective tax rates provide a useful measure of class differences in total tax burden. Often effective tax rates are computed for each income **quintile** and the top 1% of households. See Chapter 4.

Elite   A top-ranked group, especially one that exercises **power** by virtue of organizational position. For example, the military elite and the corporate elite. Properly used as a collective noun referring to a group, rather than to individual members of a group. An elite perspective on power makes a sharp distinction between an organized minority that rules and an unorganized majority that is ruled. See **elite cohesion, pluralist perspective,** and Chapter 8.

Elite Cohesion   The degree to which members of a hypothesized **elite** band together in pursuit of common objectives and in opposition to other groups. The greater the degree of elite cohesion, the more likely it will be able to impose its will on others. See Chapter 8.

Elite Perspective   See **pluralist perspective.**

Endogamy   Marriage between partners drawn from the same social group. Class endogamy tends to preserve class differences in customs, values, and attitudes.

Entitlement   A government benefit program, available to all who meet specified prerequisites and supported by open-ended government funding. Entitlements can either be **means-tested,** such as the food stamp program, or available to people of varied income levels, such as Social Security.

Establishment, the   An informal network of wealthy, powerful men drawn from the upper class. During the first three-quarters of the twentieth century, members frequently filled important government positions and influenced national policy, especially in international and economic affairs.

Family   In studies of income, two or more related individuals residing together. The most common type of **household.**

Gilbert-Kahl Model of the Class Structure   The authors' model of the American class structure. Based on economic distinctions, particularly source of **income** (notably, assets, jobs, and government transfers) and occupation, rather than on prestige distinctions. The model divides the American class system into six classes: the **capitalist class,** the **upper-middle class,** the **middle class,** the **working class,** the **working poor,** and the **underclass.** See Figure 1.1 (p. 13) and Chapter 11.

Government Transfers   Payments to individuals, such as Social Security, veterans benefits, and public assistance, that are not directly in exchange for goods or services provided.

Gross Assets   A measure of **wealth** that is equal to the total value of the assets someone owns.

Household   In studies of income, a domestic unit consisting of **families,** individuals residing alone, or unrelated persons residing together.

Ideology   Term used by Marx to refer to the dominant ideas of a society, especially those that justify the *status quo,* along with the privileges and power of the ruling class. Ideology may be explicitly political or subtly, even unconsciously, embedded in conceptions of religion, the family, education, law, and so forth. Marx argued that a society's ideology is controlled by the dominant class through its power over the institutions that create and disseminate ideas, such as schools, mass media, churches, and courts.

Income   The inflow of money over a *period of time* (e.g., $500 a week or $50,000 in 1999). Distinguished from **wealth,** which refers to assets owned at a *point in time.* The primary source of income for most households is job **earnings.** Others include **government transfers,** such as Social Security benefits and public assistance; interest on bank accounts or bonds; dividends from corporate stock shares; profits from a business or professional practice; and profits from the sale of assets. See Chapter 4.

Industrial Society   See **postindustrial society.**

Inflow Mobility Table   See **outflow mobility table.**

In-Kind Benefits   Noncash government benefits, such as health care, food stamps, and subsidized housing.

Intergenerational Mobility   See **social mobility.**

Investor Class   See **wealth classes.**

Joint Marital Relationships   Marital relationships that focus on companionship and deemphasize the sexual division of labor. Husbands and wives in joint relationships share the planning of family affairs, carry out many household duties interchangeably, and value common leisure activities. More frequent at higher class levels. Contrasted with **segregated marital relationships.**

Life Chances   Aspects of an individual's future possibilities that are shaped by class membership, from the infant's chances for decent nutrition to the adult's opportunities for worldly success. First used by Max Weber to emphasize the extent to which economic position shapes each person's chance of attaining the good things in life.

Lifestyle   Introduced by Weber to describe distinctive patterns of social interaction, leisure, consumption, dress, language, and so on, associated with a social group—in particular, a prestige class or, in Weber's terminology, a "status group."

Lower Classes   The **working poor** and **underclass,** as defined in this book.

Majority Classes   The **middle class** and the **working class,** as defined in this book.

Mean   The mathematical average (sum of all values divided by the number of cases). Along with **median,** a measure of central tendency used to compare groups or measure change in a particular group over time.

Means of Production   In Marx, the implements and physical structures that are necessary for production such as land, machines, mines, or factories. Marx argued that a person's social relationship to the means of production defines his or her class position. He defined the **bourgeoisie** as the class that owns the means of production in a capitalist society and the **proletariat** as the class of workers compelled to sell their labor to the owners in order to survive.

Means-Tested Programs   Government benefits only available to people with incomes below a specified level. See **entitlement.**

Median   One way of measuring the average. More precisely, the midpoint in any distribution dividing the top 50% from the bottom 50%. Often used to make income comparisons among groups or over time periods. Unlike the **mean,** the median income figure is not distorted by extreme incomes at the high end of the distribution.

Middle Class   One of the two largest classes in the **Gilbert-Kahl model.** Located below the **upper-middle class** and above the **working class.** Composed of lower managers, semiprofessionals, crafts workers, foremen, and nonretail sales people. A higher level of skill or knowledge and independence required on the job distinguishes the middle class from the working class below. More loosely, middle class refers to the upper part of the class hierarchy in contrast to the working class or lower part.

Minimum Wage   The legal minimum that employers in most fields can pay. Important because the earnings of low-wage workers who receive more than the minimum are influenced by it. Set every few years by federal legislation, but has tended, especially since the 1980s, to lag behind inflation and thus lose purchasing power.

Mode of Production   Marx's term for a society's basic socioeconomic system. Encompasses both the technology with which the society meets its economic needs and the social organization of production. Feudalism and capitalism are distinctive modes of production.

Multiple Causal Pathways   Causation through two or more channels. For example, father's occupation may influence daughter's education, which in turn influences her own occupation. But father's occupation may also influence daughter's occupation later on, in a separate way, when the father directly influences the daughter's chances of finding a good job.

Nearly Propertyless Class   See **wealth classes.**

Nest-Egg Class   See **wealth classes.**

Net Worth   See **wealth.**

New Middle Class/Old Middle Class   C. Wright Mills's distinction between an old middle class of small entrepreneurs, farmers, shopkeepers, and independent professionals and a new middle class of salaried white-collar workers, including managers, employed professionals, office workers, and salespeople. The first is distinguished by its dependence on entrepreneurial property, the second by its dependence on salaried employment. They have long coexisted, but the first has been shrinking since the late nineteenth century, while the second has been growing. Changes in the economy after the initial stage of industrialization favor the growth of the new middle class. See Chapter 3.

Occupation   A social role that describes the major work that a person does to earn a living. An individual's occupation or that of a family's main income earner is a key determinant of class position. See Chapter 3.

Occupational Prestige   The status or respect accorded an occupation. Measured by occupational prestige scores derived from opinion surveys. Occupational prestige influences personal **prestige,** especially when people do not have detailed knowledge of each other's income, family background, lifestyle, associations, and so on. See **socioeconomic status** and Chapter 2.

Occupational Structure   The proportional distribution of workers in the different occupational categories. The occupational structure shifts with changes in the economy. For example, when the United States changed from an agricultural society to an industrial society, the proportion of farmers decreased and the proportion of **operatives** increased. See Chapter 3, especially Table 3.2 (p. 49).

125% of the Poverty Line   A broader measure of poverty than the official federal **poverty standard,** also issued by the Census Bureau. Based on poverty thresholds 25% higher than the official standard. A higher poverty line means more households are classified as poor.

Operatives   An occupational category composed of semiskilled and unskilled workers who operate machines, including assembly-line workers in manufacturing, butchers, gas station attendants, and truck drivers. The proportion of operatives increased as the country industrialized and then decreased during the **postindustrial** period.

Outflow Mobility Table/Inflow Mobility Table   Cross tabulations of father's occupation in the past and son's or daughter's current occupation. The outflow table conceptually groups fathers into occupational categories and examines the percentage of occupational distributions of their sons or daughters. For example, such a table might include a row showing the occupations of all sons or daughters of upper–white-collar men. An outflow table suggests the extent to which careers are influenced by class background. The inflow table groups sons or daughters into occupational categories and examines the corresponding percentage distributions of their fathers. Inflow tables indicate the diversity of class origins among people in each occupational category. The outflow table asks, "Where did they go?" The inflow table asks, "Where did they come from?" Outflow tables are typically percentaged across the rows; inflow tables, down the columns. See Chapter 6, especially Tables 6.1 (p. 124) and 6.2 (p. 125).

Outsourcing   Corporate strategy to lower labor costs by purchasing services, components, or finished products from low-wage companies at home and abroad. Especially common among large American companies since the 1980s.

Path Analysis   A method of causal analysis that uses formal path diagrams and related calculations to sort out the influences of a series of variables on one another and on an ultimate dependent variable. Used especially to examine the factors responsible for career success or failure. Such a model might, for example, be used to explore how a variety of social background variables and education influence each other and ultimate occupational achievement. See Chapter 7.

Pink-Collar Occupations   Occupations that are largely female. These fields typically offer lower pay, less prestige, and slimmer opportunities for advancement than male-dominated occupations requiring similar levels of education and training. Examples: secretaries, cashiers, hairdressers, nurses, and elementary school teachers.

Pluralist Perspective   A theoretical perspective that regards **power** as typically diffused rather than concentrated. Pluralists find multiple bases of power representing the interests of competing groups, such that no powerful minority can easily impose its will. Contrasted with the elite perspective that emphasizes the power of a ruling minority, and the class perspective, which associates power with a dominant social class, such as the capitalist class. See **elite** and Chapter 8.

Political Action Committee (PAC)   A group formed, under the provisions of the campaign finance laws, to raise and contribute money to the campaigns of political candidates.

Postindustrial Society   As defined in this book, a society in which most workers are employed in the **service-producing sectors,** like law, education, and retail, rather than the goods-producing sectors of the economy. The United States has evolved from an *agricultural society,* in which most workers were farmers (1776 to 1900); to an *industrial society,* in which the majority of workers were employed in manufacturing or related fields such as mining, transportation and utilities (1900 to 1970); to a *postindustrial society* (since 1970). See Chapter 3, especially Figure 3.2 (p. 52).

Poverty Line   See **poverty threshold.**

Poverty Rate   The percentage of people, families, or households who are poor by some standard (usually the official federal standard). See other **poverty** entries and Chapter 10.

Poverty Standard, Federal (Official)   Standard established in the 1960s to measure poverty rates and numbers of poor people or poor households. Originally based on the cost of feeding an average family and the proportion of the family budget devoted to nonfood expenses. Takes household size into account. The federal measure is an **absolute poverty standard.** It is adjusted annually for inflation, but not for changes in the standard of living. See Chapter 10.

Poverty Standards, Absolute and Relative   Absolute poverty standards define poverty as not having enough food, adequate housing, and so on. They represent a fixed material standard. Relative standards define poverty as having significantly less than the average member of the community. They change over time as the community standard of living changes. The official poverty measure is an absolute standard. A commonly proposed relative standard is half the **median** income. Critics of the official standard, who favor a relative poverty standard, see absolute standards as irrelevant to an affluent society with rising living standards, like the United States. See **poverty standard, federal,** and Chapter 10.

Poverty Threshold   The income below which a household of a given size is classified as poor. Federal poverty statistics are calculated using a series of such thresholds, which are based on the **federal poverty standard,** periodically adjusted for inflation. Also referred to as the poverty line.

Power   The capacity of individuals or groups to carry out their will even over the opposition of others, especially in broad political and economic contexts. For example, the power of the capitalist class over national economic priorities. See Chapter 8.

Prestige   Social esteem or honor. Expressed in attitudes of respect or deference in social interaction. Sometimes used interchangeably with **status.** When a group of people or families in a community share a common position of prestige, they may be described as a prestige class. See Chapter 2.

Privileged Classes   The **capitalist class** and **upper-middle class,** as defined in this book.

Progressive Tax   A tax that takes higher proportions of income at higher income levels. The federal income tax, with its escalating marginal tax rates, is the most important progressive tax. See **regressive tax.**

Proletariat   Marx's term for the working class, defined as people who must sell their labor power to business owners to survive in a capitalist society.

Quintile   A ranked fifth of the population being studied. Comparisons of earnings, income, and wealth are often made among quintiles of households, ranked from the richest quintile to the poorest.

Ranking Consistency   The degree of community consensus about rankings in a hierarchy of individuals, families, or occupations. Studies show greater consensus about rankings than about division of the hierarchy into classes. See **cutting consistency** and Chapter 2.

Real Income (also real earnings, wages, etc.)   Income adjusted for inflation so that comparisons over long periods can be made. Always denominated in dollars of a specific year. Similar to conversion of prices from one national currency into another. For example, incomes from the 1970s restated in 2000 dollars would be several times higher, while representing the same purchasing power.

Regressive Tax   A tax that takes higher proportions of income at lower income levels. The retail sales tax, though charged at a flat rate, is a regressive tax because low-income households, unlike high-income households, spend virtually all their income on retail items. See **progressive tax.**

Relative Poverty Standard   See **poverty standards.**

Risk of Poverty   The chance of being poor if you belong to a specified social group, expressed as the percentage of people in that social group who fall below the poverty line. See **poverty** entries and Figure 10.2 (p. 212).

Segregated Marital Relationships   Marital relationship in which there is clear differentiation of concerns and responsibilities that minimizes the husband's involvement with household matters and the wife's with the world of the husband's work. Segregated marital relationships are most common at lower class levels. Contrasted with **joint marital relationships.**

Service Sectors (also service-producing sectors)   Economic sectors, including retail, finance, law, hotels, health care, and education, that produce services. Contrasted with goods-producing sectors like agriculture, manufacturing, and construction. The service-producing sectors grow rapidly in a **postindustrial society.**

Service Workers   The occupational category of workers who provide a service, including waiters, janitors, child care workers, domestic servants, police and firefighters—generally, but not always, low-skill, low-pay occupations. Employment in service occupations expands in a **postindustrial society.** Not all service workers are employed in the **service sector** of the economy. (There are janitors in manufacturing, for example.) Not all service sector jobs are performed by service workers. (Doctors and lawyers work in service sectors.)

Social Capital.   See **Capital.**

Social Class   A large group of families approximately equal in rank and differentiated from other families with regard to characteristics such as **occupation, prestige,** or **wealth.** See Chapter 1.

Social Clique   According to W. Lloyd Warner, an intimate nonkin group with no more than 30 members. Warner found that most social cliques are composed of people of the same or adjacent classes.

Social Mobility   The extent to which people move up or down in the class system, typically measured by occupational status. Intergenerational mobility is the movement of individuals relative to their parents' position. Social succession is the inheritance of parents' position. Intragenerational mobility is movement in the course of an individual's own career. See Chapters 6 and 7.

Social Status   See **status.**

Social Strata   See **strata, social.**

Social Stratification   Ranking of individuals or families based on characteristics such as **occupation, income, wealth,** and social **prestige.** The hierarchy may be continuous or divided into a series of discrete **social classes,** consisting of people of roughly equal rank. Thus, class is a special case of stratification.

Social Succession   See **social mobility.**

Socialization   The process through which people learn the skills, attitudes, and customs needed to participate in the life of the community. Class-specific socialization reinforces distinctive class attitudes and lifestyles and encourages individuals to assume the class position of their parents. See Chapter 5.

Socioeconomic Status (SES)   Social standing or prestige, especially as measured by the occupational prestige scores.

Soft Money   Political contributions not subject to federal regulation because they are putatively for party-building activities, voter registration, and so on, but not for the campaigns of individual federal candidates. In practice, the distinction is almost meaningless. With soft money, individuals, corporations, unions, and other organizations have been able to make political contributions on a scale (hundreds of thousands of dollars) that would normally be prohibited by federal law. A 2002 law attempted to place strict limits on the use of soft money. See Chapter 9.

Status   Synonym for social **prestige**. Also, used more generally in sociology to refer to any distinctive social condition or position. Weber wrote about "status groups," which he described as social communities of people who share a common position of social honor (or dishonor). Weber distinguished status groups from economically defined classes. His use of the concept established the idea of separate social and economic dimensions of stratification systems. See Chapter 1.

Strata, Social   Levels in a stratified hierarchy, especially those based on prestige.

Structural Mobility   Mobility made possible by changes in the occupational structure. A relative expansion of middle- or upper-level jobs, for example, would allow upward mobility from lower positions. See **circulation mobility** and Chapter 6.

Superstructure   Marx's term for the dominant political institutions and **ideology** of a society. The means by which the ruling class controls a society. Marx distinguished between the superstructure and the socioeconomic base or foundation of a society.

Temporary Aid to Needy Families (TANF)   Replaced **AFDC** in 1996 as the income assistance program for impoverished children and their families. Different from AFDC in that TANF is not an **entitlement**. Under TANF, families generally may not receive aid for more than a lifetime total of 5 years, and adults benefiting from TANF must start some sort of work within 2 years after assistance begins. See Chapter 10.

Two-Tier Wage Systems   The practice of paying new employees less than employees with longer job tenure. Used by some American corporations to reduce labor costs, especially since the 1980s.

Underclass   The bottom class in the **Gilbert-Kahl model**. Low-income families with a tenuous relationship to the job market, whose incomes often depend on government programs. The term is sometimes used in a negative sense to refer to a class of people who are impoverished and mired in habits and circumstances that prevent them from ever joining the mainstream. See Chapter 10.

Upper–Middle Class   In the **Gilbert-Kahl model,** the class below the capitalist class and above the middle class, consisting of well-paid, university-trained managers and professionals.

Variance Explained   An indicator of the accuracy with which a series of antecedent variables can predict the values of a dependent variable. For example, how accurately can a son's occupation be predicted from knowledge of his education, his father's occupation, and his father's education? Stated as a percentage of total variance. Because outcomes are always affected by chance factors that are difficult to take into account and our measurements are never perfectly accurate, the variance explained seldom exceeds 50%. The difference between 100% and the variance explained is the unexplained variance.

Wage-Setting Institutions   Institutions, such as labor unions, internal labor markets within corporations, and minimum wage legislation, that shield workers from market forces by influencing decisions that would otherwise be determined by supply and demand.

Wealth   The value of assets owned by an individual or family at a *point in time* (e.g., $150,000 on Dec. 31, 2001). Distinguished from **income,** which refers to value received over a *period of time.* Typical forms of wealth are homes, automobiles, bank accounts, and business or financial assets, including commercial real estate, small enterprises, stocks, and bonds. An important distinction is made between wealth held for personal use, such as a home, and wealth held in the form of income-producing assets such as stocks or commercial real estate. Wealth is typically measured as net worth, the value of assets owned less the amount of debt owed.

Wealth Classes   Three classes, conceptualized by the authors, that describe the distribution of wealth in the United States. The **nearly propertyless class,** about 40% of the population, has little or even negative net worth. The **nest-egg class,** about 50% of the population, typically has significant equity in homes and cars, plus a thin cushion of interest-earning assets, such as bank accounts, and modest holdings of stocks or mutual funds. The **investor class,** 10% of the population, owns most of the privately held investment assets and typically controls large diversified portfolios.

White-Collar Workers   Office workers, including the U.S. Census categories of managers, professionals, clerical workers, and sales workers. Sometimes used to refer to the **middle class.** Distinguished from **blue-collar** and **service workers.**

Winner-Take-All Markets   Term used by economists Frank and Cook (1995) to describe markets in which the rewards are heavily concentrated in the hands of a few top performers who are just slightly better than their closest competitors. Common in entertainment and professional sports but, according to these authors, also increasing across the economy. Important example: rise of corporate CEO compensation to multimillion-dollar levels in recent years.

Working Class   One of the two largest classes in the **Gilbert-Kahl model.** Located below the middle class and above the working poor. Composed of low-skill manual workers, clerical workers, and retail salespeople. Commonly, the term is used to refer to the lower portion of the class structure or all blue-collar workers.

Working Poor   In the **Gilbert-Kahl model,** the class below the working class and above the underclass, consisting of people who hold low-wage, low-skill, often insecure jobs typically involving menial blue-collar, sales, or service work.

Working Rich   A small stratum at the top of the **upper-middle class,** consisting of very successful professionals (notably, lawyers, doctors, dentists), small business owners, and ranking (but not top) corporate executives, all with incomes in the hundreds of thousands.

Bibliography

Abramson, Paul R., John H. Aldrich, and David W. Rohde. 1995. *Change and Continuity in the 1992 Elections.* Rev. ed. Washington, DC: Congressional Quarterly Press.

Acker, Joan 2006. *Class Questions: Feminist Answers.* Lanham, MD: Rowman & Littlefield.

Aldrich, Nelson W., Jr. 1988. *Old Money: The Mythology of America's Upper Class.* New York: Vintage.

Alexander, Herbert E. and Brian A. Haggerty. 1987. *Financing the 1984 Elections.* Lexington, MA: Heath.

Allan, Graham. 1989. *Friendship: Developing a Sociological Perspective.* Boulder, CO: Westview.

Allen, Frederick Lewis. 1952. *The Big Change.* New York: Harper & Row.

Alvarez, Louis and Andrew Kolker. 1999. *People Like Us: Social Class in America* [Video]. Public Broadcasting Service.

Anderson, Dewey and Percy Davidson. 1943. *Ballots and the Democratic Class Struggle.* Palo Alto, CA: Stanford University Press.

Anderson, Elijah. 1992. *Streetwise: Race, Class, and Change in an Urban Community.* Chicago: University of Chicago Press.

————. 1999. *Code of the Street: Decency, Violence, and the Moral Life of the Inner City.* New York: Norton.

Anderson, Martin. 1978. *Welfare: The Political Economy of Welfare Reform in the United States.* Palo Alto, CA: Hoover Press.

Argyle, Michael 1994. *The Psychology of Social Class.* New York: Routledge.

Arrow, Kenneth, Samuel Bowles, and Steven Durlauf, eds. 2000. *Meritocracy and Economic Inequality.* Princeton, NJ: Princeton University Press.

Auletta, Ken. 1982. *The Underclass.* New York: Random House.

Bachrach, Peter and Morton S. Baratz. 1974. *Power and Poverty: Theory and Practice.* New York: Oxford University Press.

Baltzell, E. Digby. 1958. *Philadelphia Gentlemen.* New York: Free Press.

Bane, Mary Jo and David Ellwood 1994. *Welfare Realities: From Rhetoric to Reform.* Cambridge, MA: Harvard University Press.

Barnouw, Erik. 1978. *The Sponsor: Notes on a Modern Potentate.* New York: Oxford University Press.

Beeghley, Leonard. 1996. *The Structure of Social Stratification in the United States.* 2nd ed. Needham Heights, MA: Simon & Schuster.

Beeghley, Leonard and John K. Cochran. 1988. "Class Identification and Gender Role Norms Among Employed Married Women." *Journal of Marriage and the Family* 50:546–566.

Bell, Daniel. 1976. *The Coming of the Post-Industrial Society.* New York: Basic Books.

Bendix, Reinhard and Seymour Martin Lipset, eds. 1953. *Class, Status, and Power.* 1st ed. Glencoe, IL: Free Press.

———. 1966. *Class, Status and Power: Social Stratification in Comparative Perspective.* 2nd ed. New York: Free Press.

Berg, Ivar. 1970. *Education and Jobs: The Great Training Robbery.* New York: Praeger.

Berle, Adolf A., Jr. and Gardiner C. Means. 1932. *The Modern Corporation and Private Property.* New York: Commerce Clearing House.

Berman, Paul. 1991. "A Union Man From Harvard." *New York Times Book Review.* August 11.

Bianchi, Suzanne. 1995. "Changing Economic Roles of Women and Men." In *State of the Union: America in the 1990s, Volume One: Economic Trends,* edited by R. Farley. New York: Russell Sage Foundation.

Birmingham, Stephen. 1987. *American's Secret Aristocracy.* Boston: Little, Brown.

Birnbach, Lisa, ed. 1980. *The Official Preppy Handbook.* New York: Workman.

Blank, Rebecca M. 2007. "What We Know, What We Don't Know, and What We Need to Know About Welfare Reform." Paper presented at the conference, "Ten Years After: Evaluating the Long-Term Effects of Welfare Reform on Children, Families, Work and Welfare." University of Kentucky, Center for Poverty Research.

Blau, Francine. 1984. "Women in the Labor Force: An Overview." In *Women: A Feminist Perspective,* edited by J. Freeman. Palo Alto, CA: Mayfield.

Blau, Peter M. and Otis Dudley Duncan. 1967. *The American Occupational Structure.* New York: Wiley.

Bloch, Fred. 1977. "The Ruling Class Does Not Rule: Notes on the Marxist Theory of the State." *Socialist Revolution* 7(1):6–28.

Blumberg, Paul M. and P. W. Paul. 1975. "Continuities and Discontinuities in Upper-Class Marriages." *Journal of Marriage and the Family* 37:63–77.

Blumenthal, Sydney. 1986. *The Rise of the Counter Establishment.* New York: Times Books.

Blumenthal, Sydney and Thomas Byrne Edsall, eds. 1988. *The Reagan Legacy.* New York: Pantheon.

Boston, Thomas D. 1988. *Race, Class, and Conservatism.* Cambridge, MA: Unwin Hyman.

Bott, Elizabeth. 1954. "The Concept of Class as a Reference Group." *Human Relations* 7:259–286.

———. 1964. *Family and Social Network.* London: Tavistock.

Bottomore, Tom. 1966. *Elites in Modern Society.* New York: Pantheon.

Bound, John and Richard B. Freeman. 1992. "What Went Wrong? The Erosion of Relative Earnings and Employment Among Young Black Men in the 1980s." *Quarterly Journal of Economics* 107:201–231.

Bourdieu, Pierre. 1984. *Distinction: A Social Critique of the Judgment of Taste.* Cambridge, MA: Harvard University Press.

———. 1986. The Forms of Capital. In *Handbook of Theory and Research for the Sociology of Education.* Edited by John Richardson. New York: Greenwood Press.

Bowles, Samuel and Herbert Gintis. 1976. *Schooling in Capitalist America.* New York: Basic Books.

Bowles, Samuel, Herbert Gintis, and Melissa Osborne Groves, eds. 2005. *Unequal Chances: Family Background and Economic Success.* New York: Russell Sage Foundation.

Bowser, Benjamin P. 2007. *The Black Middle Class: Social Mobility—and Vulnerability.* Boulder, CO: Lynne Rienner.

Boyer, Richard and Herbert Morais. 1975. *Labor's Untold Story.* 3rd ed. New York: United Electrical Workers.

Bradburn, Norman. 1969. *The Structure of Psychological Well-Being.* Chicago: Aldine.

Bradbury, Katherine and Jane Katz. 2002. "Issues in Economics." *Regional Review* Q4:2–5.

Braverman, Harry. 1974. *Labor and Monopoly Capital.* New York: Monthly Review Press.

Brewer, Mark D. and Jeffrey M. Stonecash. 2007. *Split: Class and Cultural Divides in American Politics.* Washington, DC: CQ Press.

Brody, David. 1993. *Workers in Industrial America: Essays on the 20th Century Struggle.* 2nd ed. New York: Oxford University Press.

Bronfenbrenner, Urie. 1966. "Socialization Through Time and Space." In *Class, Status, and Power,* 2nd ed., edited by R. Bendix and S. M. Lipset. New York: Free Press.

Brooks, David. 2000. *Bobos in Paradise: The New Upper Class and How They Got There.* New York: Simon & Schuster.

Brooks, Thomas. 1971. *Toil and Trouble: A History of American Labor.* 2nd ed. New York: Dell.

Brown, Clifford et al. 1995. *Serious Money: Fundraising and Contributing in Presidential Nominational Campaigns.* New York: Cambridge University Press.

Brunner, Borgna, ed. 2001. *Time Almanac 2002.* Boston: Information Please.

Bucks, Brian, Arthur Kennickell, and Kevin Moore 2006. "Recent Changes in U.S. Family Finances: Evidence From the 2001 and 2004 Survey of Consumer Finances." *Federal Reserve Bulletin* 92:A1–A38.

Burch, Philip H., Jr. 1980. *Elites in American History: The New Deal to the Carter Administration.* New York: Holmes and Meier.

Burke, Vee 2001. *Welfare Reform: TANF Trends and Data. CRS Report for Congress.* Washington, DC: Congressional Research Service.

Burke, Vee et al. 2001. *Welfare Reform Briefing Book.* Washington, DC: Congressional Research Service.

Burnham, James. 1941. *The Managerial Revolution.* New York: John Day.

Burtless, Gary. 1987. "Inequality in America: Where Do We Stand?" *Brookings Review* 5 (Summer):9–16.

———. 1990. *A Future of Lousy Jobs.* Washington, DC: Brookings Institution.

———. 1995. "International Trade and the Rise in Earnings Inequality." *Journal of Economic Literature* 32:800–816.

Burtless, Gary, R. Kent Weaver, and Joshua Wiener. 1997. "The Future of the Social Safety Net." In *Setting National Priorities: Budget Choices for the Next Century,* edited by R. Reishauer. Washington, DC: Brookings Institution.

BusinessWeek. 1996. "Special Report: How High Can CEO Pay Go?" April 22.

BusinessWeek. 1997. "Executive Pay: Special Report." April 21.

BusinessWeek. 2000. "Special Report: Executive Pay." April 17.

BusinessWeek. 2001. "Special Report: Executive Pay." April 16.

Cameron, Juan. 1978. "Small Business Trips Big Labor." *Fortune* 98 (July):80–82.

Campbell, Angus, Gerald Gurin, and Warren E. Miller. 1954. *The Voter Decides.* Evanston, IL: Row, Peterson.

Cancian, Maria et al. 1993. "Working Wives and Family Income Inequality Among Married Couples." In *Uneven Tides: Rising Inequality in America,* edited by S. Danziger and P. Gottshalk. New York: Russell Sage Foundation.

Cantril, Hadley. 1951. *Public Opinion.* Princeton, NJ: Princeton University Press.

Caplow, Theodore. 1980. "Middletown Fifty Years After." *Contemporary Sociology* 9:46–50.

Caplow, Theodore and Bruce Chadwick. 1979. "Inequality and Life Styles in Middletown, 1920–1978." *Social Science Quarterly* 60:366–368.

Caplow, Theodore et al. 1985. *Middletown Families: Fifty Years of Change and Continuity.* Minneapolis: University of Minnesota Press.

Card, David and Alan B. Krueger. 1992. "School Quality and Black–White Relative Earnings: A Direct Assessment." *Quarterly Journal of Economics* 107:151–200.

Cashell, Brian. 2000. The Distribution of Household Wealth in the United States. *CRS Report for Congress.* Congressional Research Service. March 16.

Center on Budget and Policy Priorities. 2001. "Poverty Rates Fell in 2000 as Unemployment Reached 31-Year Low." Press Release, September 26. Washington, DC: Author.

Centers, Richard. 1949. *The Psychology of Social Classes: A Study of Class Consciousness.* Princeton, NJ: Princeton University Press.

Charles, Kerwin Kofi and Erik Hurst. 2003. "The Correlation of Wealth Across Generations." *The Journal of Political Economy* 111:1155–1182.

Charlot, Monica. 1985. "The Ethnic Minorities Vote." In *Britain at the Polls,* edited by A. Ranney. Washington, DC: American Enterprise Institute.

Citizens for Tax Justice and the Institute on Taxation and Economic Policy. 1996. *Who Pays? A Distributional Analysis of the Tax Systems in All 50 States.* Washington, DC: Author.

Cohen, Jere. 1979. "Socio-Economic Status and High School Friendship Choice: Elmtown's Youth Revisited." *Social Networks* 2:65–74.

Colclough, Glenna, and E. M. Beck. 1986. "The American Educational Structure and the Reproduction of Social Class." *Sociological Inquiry* 56:456–476.

Coleman, Richard P. and Lee Rainwater, with Kent A. McClelland. 1978. *Social Standing in America: New Dimensions of Class.* New York: Basic Books.

Collier, Peter and David Horowitz. 1976. *The Rockefellers: An American Dynasty.* New York: Holt, Rinehart & Winston.

Collins, Chuck and Felice Yeskel, with United for a Fair Economy and Class Action. 2005. *Economic Apartheid in America: A Primer on Economic Inequality & Insecurity.* Revised ed. New York: New Press.

Congressional Budget Office. 2001. "Effective Federal Tax Rates: 1979–1997."

Congressional Budget Office. 2006. "Historical Effective Federal Tax Rates: 1979 to 2004." Available at http://www.cbo.gov/ftpdocs/77xx/doc7718/EffectiveTaxRates.pdf

Congressional Quarterly. 1976. *Guide to Congress.* 2nd ed. Washington, DC: Congressional Quarterly Press.

Congressional Quarterly. 1980. "Democrats May Lose Edge in Contributions From PACs." 38:3405–3409.

———. 1996. "Welfare Overhaul Law." (Sept. 21):2696–2705.

Conley, Dalton. 1999. *Being Black, Living in the Red: Race, Wealth, and Social Policy in America.* Berkeley: University of California Press

Cookson, Peter and Caroline Hodges Persell. 1985. *Preparing for Power: America's Elite Boarding Schools.* New York: Basic Books.

Corcoran, M. 1995. "Rags to Rags: Poverty and Mobility in the United States." *Annual Review of Sociology* 21:237–267.

Corey, Lewis. 1935. *The Crisis of the Middle Class.* New York: Covici-Friede.

———. 1953. "Problems of the Peace: The Middle Class." In *Class, Status, and Power,* edited by R. Bendix and S. M. Lipset. Glencoe, IL: Free Press.

Corrado, Anthony, Thomas E. Mann, Daniel R. Ortiz, and Trevor Potter. 2005. *The New Campaign Finance Sourcebook.* Washington, DC: Brookings Institution Press.

Coser, Lewis. 1978. *Masters of Sociological Thought.* 2nd ed. New York: Harcourt Brace Jovanovich.

Coverman, Shelly. 1988. "Sociological Explanations of the Male–Female Wage Gap." In *Women Working: Theories and Facts in Perspective,* 2nd ed., edited by A. Stromberg and S. Harkess. Mountain View, CA: Mayfield.

Coxon, Anthony and Charles Jones. 1978. *The Images of Occupational Prestige.* New York: St. Martin's.

Crompton, Rosemary. 1993. *Class and Stratification: An Introduction to Current Debates.* Cambridge, MA: Polity.

———. 1998. *Class and Stratification: An Introduction to Current Debates.* 2nd ed. Cambridge, MA: Polity.

Croteau, David. 1995. *Politics and the Class Divide: Working People and the Middle-Class Left.* Philadelphia: Temple University Press.

Cruciano, Therese. 1996. "Individual Tax Returns: Preliminary Data, 1994." *SOI Bulletin: A Quarterly Statistics of Income Report* 15(Spring):18–24.

Current Biography. 1982–1988. New York: W. W. Wilson.

Curtis, Richard F. and Elton F. Jackson. 1977. *Inequality in American Communities.* New York: Academic.

Daalder, Hans and Ruud Koole. 1988. "Liberal Parties in the Netherlands." Pp. 151–177 in *Liberal Parties in Western Europe* edited by E. Kirchner. Cambridge, UK: Cambridge University Press.

Dahl, Robert A. 1961. *Who Governs? Democracy and Power in an American City.* New Haven, CT: Yale University Press.

———. 1967. *Pluralist Democracy in the United States.* Chicago: Rand McNally.

Danziger, Sheldon and Peter Gottschalk. 1995. *America Unequal.* New York: Russell Sage Foundation.

Davidson, James D., Ralph E. Pyle, and David V. Reyes. 1995. "Persistence and Change in the Protestant Establishment, 1930–1992." *Social Forces* 74(1):157–175.

Davis, Allison, Burleigh B. Gardner, and Mary R. Gardner. 1941. *Deep South: A Social-Anthropological Study of Caste and Class.* Chicago: University of Chicago Press.

Davis, James Allen and Tom Smith. 1990. *General Social Surveys, 1972–1990* (machine-readable data file). Chicago: National Opinion Research Center.

Davis, Mike. 1980. "The Barren Marriage of American Labour and the Democratic Party." *New Left Review* 124:45–84.

DeCarlo, Scott. 2007. "Big Paychecks." *Forbes* May 21.

De Luca, Rita Caccamo. 2001. *Back to Middletown: Three Generations of Sociological Reflections.* Palo Alto, CA: Stanford University Press.

Demerath, N. J., III. 1965. *Social Class in American Protestantism.* Chicago: Rand McNally.

Dent, David. 1992. "The New Black Suburbs." *New York Times Magazine.* June 14.

DeParle, Jason. 1996. "Welfare: Progress Hijacked." *New York Times Magazine.* December 8.

———. 2004. *American Dream: Three Women, Ten Kids and a Nation's Drive to End Welfare.* New York: Penguin.

DeStefano, Linda. 1990. "Pressures of Modern Life Bring Increased Importance to Friendship." *Gallup Poll Monthly* (March):24–33.

Dinitz, Simon, Franklin Banks, and Benjamin Pasamanick. 1960. "Mate Selection and Social Class: Change in the Past Quarter Century." *Marriage and Family Living* 22:348–351.

Domhoff, G. William. 1967. *Who Rules America?* Englewood Cliffs, NJ: Prentice-Hall.

———. 1970. *The Higher Circles: The Governing Class in America.* Englewood Cliffs, NJ: Prentice Hall.

———. 1974. *The Bohemian Grove and Other Retreats.* New York: Harper & Row.

———. 1975. "Social Clubs, Policy Planning Groups, and Corporations." *Insurgent Sociologist* 5(3):173–195.

———. 2006. *Who Rules America?* 5th ed. Boston: McGraw-Hill.

Domhoff, G. William and Hoyt B. Ballard, eds. 1968. *C. Wright Mills and the Power Elite.* Boston: Beacon.

Dotson, Floyd. 1950. "The Associations of Urban Workers." Unpublished doctoral thesis, Yale University.

Dubofsky, Melvyn. 1980. "The Legacy of the New Deal." *Executive* 6 (Spring):8–10.

Duncan, Gregg J. et al. 1984. *Years of Poverty, Years of Plenty: The Changing Economic Fortunes of American Workers and Families.* Ann Arbor: Institute for Social Research, University of Michigan.

Duncan, Gregg and Willard Rodgers. 1989. "Has Poverty Become More Persistent?" Institute for Social Research, University of Michigan.

Duncan, Otis Dudley. 1961. "A Socio-Economic Index for All Occupations," and "Properties and Characteristics of the Socioeconomic Index." In *Occupations and Social Status,* edited by A. Reiss. Glencoe, IL: Free Press.

———. 1966. "Methodological Issues in the Analysis of Social Mobility." In *Social Structure and Mobility in Economic Development,* edited by N. Smelser and S. M. Lipset. Chicago: Aldine.

Duncan, Otis Dudley, Archibald O. Haller, and Alejandro Portes. 1968. "Peer Influences on Aspirations: A Reinterpretation." *American Journal of Sociology* 74:119–137.

Dye, Thomas R. 1976. *Who's Running America? Institutional Leadership in the United States.* Englewood Cliffs, NJ: Prentice Hall.

———. 1979. *Who's Running America? The Carter Years.* 2nd ed. Englewood Cliffs, NJ: Prentice Hall.

———. 1995. *Who's Running America: The Clinton Years.* 6th ed. Englewood Cliffs, NJ: Prentice Hall.

———. 2002. *Who's Running America: The Bush Restoration.* 7th ed. Englewood Cliffs, NJ: Prentice Hall.

Edelman, Peter. 1997. "The Worst Thing Bill Clinton Has Done." *Atlantic Magazine.* March.

Edin, Kathryn and Maria Kefalas. 2005. *Promises I Can Keep: Why Poor Women Put Motherhood Before Marriage.* Berkeley: University of California Press.

Edin, Kathryn and Laura Lein. 1997. *Makin' Ends Meet. How Single Mothers Survive: Welfare and Low-Wage Work.* New York: Russell Sage Foundation.

Edsall, Thomas B. 1984. *The New Politics of Inequality.* New York: Norton.

Edsall, Thomas and Mary Edsall. 1991. *Chain Reaction: The Impact of Race, Rights, and Taxes on American Politics.* New York: Norton.

Edwards, Alba M. and U.S. Bureau of the Census. 1943. *U.S. Census of Population 1940: Comparative Occupational Statistics, 1870–1940.* Washington, DC: U.S. Government Printing Office.

Ehrenreich, Barbara. 1989. *Fear of Falling: The Inner Life of the Middle Class.* New York: Pantheon.

———. 2001. *Nickel and Dimed: On (Not) Getting by in America.* New York: Henry Holt.

Eismeier, Theodore and Philip Pollock. 1996. "Money in the 1994 Elections and Beyond." In *Midterm: The Elections of 1994 in Context,* edited by P. Klinkner. Boulder, CO: Westview.

Ellwood, David. 1988. *Poor Support: Poverty in the American Family.* New York: Basic Books.

Ellwood, David and Mary J. Bane. 1985. "The Impact of AFDC on Family Structure and Living Arrangements." In *Research in Labor Economics* 7, edited by R. Ehrenberg. Greenwich, CT: JAI Press.

———. 1994. *Welfare Realities: From Rhetoric to Reform.* Cambridge, MA: Harvard University Press.

Erikson, Robert and John Goldthorpe. 1992. *The Constant Flux: A Study of Class Mobility in Industrial Societies.* New York: Oxford University Press.

Farley, Reynolds. 1984. *Blacks and Whites: Narrowing the Gap?* Cambridge, MA: Harvard University Press.

Farley, Reynolds, ed. 1995a. *State of the Union: America in the 1990s, Volume 1: Economic Trends.* New York: Russell Sage Foundation.

———. 1995b. *State of the Union: America in the 1990s, Volume 2: Social Trends.* New York: Russell Sage Foundation.

Faux, Jeff. 2006. *The Global Class War: How America's Bipartisan Elite Lost Our Future and What It Will Take to Win It Back.* New York: John Wiley.

Featherman, David 1979. "Opportunities are Expanding." *Society* March/April, 4, 6–11.

Featherman, David L. and Robert M. Hauser. 1978. *Opportunity and Change.* New York: Academic.

Ferguson, Thomas. 1995. *Golden Rule: The Investment Theory of Party Competition and the Logic of Money Driven Political Systems.* Chicago: University of Chicago Press.

Fine, Michelle and Lois Weis. 1998. *The Unknown City: The Lives of Poor and Working-Class Young Adults.* Boston: Beacon.

Fisher, Claude et al. 1996. *Inequality by Design: Cracking the Bell Curve Myth.* Princeton, NJ: Princeton University Press.

Forbes. 1990. "The Forbes 400." October 22.

———. 1991. "What 800 Companies Paid Their Bosses." May 27.

———. 1996. "The Forbes 400." October 14.

———. 1998. "Family Fortunes: The 50 Wealthiest Families in America." October 12.

———. 2001a. "The Forbes 400." October 8.

———. 2001b. "The 500 Largest Private Companies." November 26.

———. 2006. "The Richest People in America." October 9.

Fortune. 1937. "The Industrial War" 14(November):105–110, 122, 156, 160, 166.

Frank, Robert. 2007. *Richistan: A Journey Through the American Wealth Boom and the Lives of the New Rich.* New York: Crown.

Frank, Robert H. and Philip J. Cook. 1995. *The Winner-Take-All Society.* New York: Simon & Schuster.

Fraser, Douglas. 1978. "UAW President Fraser Resigns from Labor-Management Group." *Radical History Review* 18(Fall):117–121.

Frears, John. 1988. "Liberalism in France." Pp. 124–150 in *Liberal Parties in Western Europe,* edited by E. Kirchner. Cambridge, UK: Cambridge University Press.

Freeman, Richard B. 1996. "Labor Market Institutions and Earnings Inequality." *New England Economic Review.* Special Issue(May/June):157–181.

———. 2004. "What, Me Vote?" In *Social Inequality,* edited by K. M. Neckerman. New York: Russell Sage Foundation.

Frieden, Jeffry A. 2006. *Global Capitalism: Its Fall and Rise in the Twentieth Century.* New York: Norton.

Fussell, Paul. 1983. *Class: A Guide Through the American Status System.* New York: Ballantine.

Galbraith, John Kenneth. 1958. *The Affluent Society.* Boston: Houghton Mifflin.

———. 1967. *The New Industrial State.* Boston: Houghton Mifflin.

Garfinkel, Irwin and Sara McLanahan. 1986. *Single Mothers and Their Children.* Washington, DC: Urban Institute.

Gecas, Viktor. 1979. "The Influence of Social Class on Socialization." In *Contemporary Theories About the Family,* Vol. I., edited by W. R. Burr et al. New York: Free Press.

Geoghegan, Thomas. 1991. *Which Side Are You On: Trying to Be for Labor When It's Flat on Its Back.* New York: Farrar, Straus & Giroux.

Giddens, Anthony. 1973. *The Class Structure of the Advanced Societies.* New York: Harper & Row.

Ginsberg, Benjamin and Martin Shefter. 1990. *Politics by Other Means: The Declining Importance of Elections in America.* New York: Basic Books.

Gittleman, Maury. 1994. "Earnings in the 1980s: An Occupational Perspective." *Monthly Labor Review* 117:16–27.

Glenn, Norval D. and Jon P. Alston. 1968. "Cultural Distances Among Occupational Categories." *American Sociological Review* 33:365–382.

———. 1975. "The Contribution of White Collars to Occupational Prestige." *Sociological Quarterly* 16:184–189.

Graetz, Michael J. and Ian Shapiro. 2005. *Death by a Thousand Cuts: The Fight Over Taxing Inherited Wealth.* Princeton, NJ: Princeton University Press.

Graham, Lawrence Otis. 1999. *Our Kind of People: Inside America's Black Upper Class.* New York: HarperCollins.

Green, Mark. 1979. *Who Runs Congress?* 3rd ed. New York: Bantam.

Greene, Bert. 1978. *Pity the Poor Rich.* Chicago: Contemporary.

Greenstone, J. David. 1977. *Labor in American Politics.* Chicago: University of Chicago Press.

Grogger, Jeffrey and Lynn A. Karoly. 2005. *Welfare Reform: Effects of a Decade of Change.* Cambridge, MA: Harvard University Press.

Grusky, David, ed. 1994. *Social Stratification: Class, Race and Gender in Sociological Perspective.* Boulder, CO: Westview.

———. 2001. *Social Stratification: Class, Race and Gender in Sociological Perspective.* 2nd ed. Boulder, CO: Westview.

Hacker, Andrew. 1997. *Money: Who Has How Much and Why?* New York: Scribner's.

Hacker, Louis. 1970. *The Course of American Economic Growth and Development.* New York: Wiley.

Halberstam, David. 1972. *The Best and the Brightest.* New York: Random House.

Halle, David. 1984. *America's Working Man: Work, Home, and Politics Among Blue-Collar Property Owners.* Chicago: University of Chicago Press.

Hamilton, Richard. 1972. *Class and Politics in the United States.* New York: Wiley.

———. 1975. *Restraining Myths: Critical Studies of United States' Social Structure and Politics.* Beverly Hills, CA: Sage.

Harrington, Michael. 1962. *The Other America: Poverty in the United States.* New York: Macmillan.

Harrison, Bennett and Barry Bluestone. 1988. *The Great U-Turn: Corporate Restructuring and the Polarizing of America.* New York: Basic Books.

Haskins, Ronald. 2006. Statement. *Hearings on 1996 Welfare Overhaul.* U.S. House of Representatives. Committee on Ways and Means. July 19.

Haveman, Robert and Barbara Wolf. 1994. *Succeeding Generations: On the Effects of Investments in Children.* New York: Russell Sage Foundation.

Herman, Edward S. 1981. *Corporate Control, Corporate Power.* New York: Cambridge University Press.

Herrstein, Richard and Charles Murray. 1994. *The Bell Curve: Intelligence and Class Structure in American Life.* New York: Free Press.

Heymann, Jody. 2000. *The Widening Gap: Why American Working Families Are in Jeopardy and What Can Be Done About It.* New York: Basic Books.

Hodge, Robert W., and Donald Treiman. 1968. "Class Identification in the United States." *American Journal of Sociology* 73:535–547.

Hodge, Robert W., Donald J. Treiman, and Peter H. Rossi. 1966. "A Comparative Study of Occupational Prestige." In *Class, Status and Power,* 2nd ed., edited by R. Bendix and S. M. Lipset. New York: Free Press.

Hodges, Harold M. 1964. *Social Stratification: Class in America.* Cambridge, MA: Schenkman.

Hoffman, Saul. 1977. "Marital Instability and the Economic Status of Women." *Demography* 14:67–76.

Hollingshead, August B. 1949. *Elmtown's Youth.* New York: Wiley.

———. 1950. "Cultural Factors in the Selection of Marriage Mates." *American Sociological Review* 15:619–627.

Hollingshead, August B. and Frederick Redlich. 1958. *Social Class and Mental Illness: A Community Study*. New York: Wiley.

Hout, Michael. 1988. "More Universalism, Less Structural Mobility: The American Occupational Structure in the 1980s." *American Journal of Sociology* 93:1358–1400.

———. 2001. "Social Mobility." In *Oxford Companion to Politics of the World*, 2nd ed., edited by J. Krieger. New York: Oxford University.

———. 2004. "How Inequality May Affect Intergenerational Mobility." In *Social Inequality*, edited by Kathryn M. Neckerman. New York: Russell Sage Foundation.

Hout, Michael et al. 1995. "Class Voting in the U.S. 1948–92." *American Sociological Review* 60:802–828.

Howe, Louise. 1977. *Pink Collar Worker: Inside the World of Woman's Work*. New York: Putnam.

Inkeles, Alex and Peter H. Rossi. 1956. "National Comparisons of Occupational Prestige." *American Journal of Sociology* 61:329–339.

Jackman, Mary. 1979. "The Subjective Meaning of Social Class Identification in the United States." *Public Opinion Quarterly* 43:443–462.

Jargowsky, Paul. 1996. "Take the Money and Run: Economic Segregation in U.S. Metropolitan Areas." *American Sociological Review* 61:984–998.

Jaynes, Gerald David and Robin M. Williams, Jr., eds. 1989. *A Common Destiny: Blacks and American Society*. Washington, DC: National Academy Press.

Jefferson, Thomas. [1821] 1944. "Autobiography." In *The Life and Selected Writings of Thomas Jefferson*, edited by A. Koch and W. Peden. New York: Modern Library.

Jencks, Christopher et al. 1972. *Inequality: A Reassessment of the Effect of Family and Schooling in America*. New York: Basic Books.

———. 1979. *Who Gets Ahead?* New York: Basic Books.

———. 1991. "Is the American Underclass Growing?" In *The Urban Underclass*, edited by C. Jencks and P. Peterson. Washington, DC: Brookings Institution.

Jencks, Christopher and Paul Peterson. 1991. *The Urban Underclass*. Washington, DC: Brookings Institution.

Johnson, Haynes, and David Broder. 1996. *The System: The American Way of Politics at the Breaking Point*. Boston: Little Brown.

Judis, John B. 1991. "Twilight of the Gods." *Wilson Quarterly* 5(Autumn):43–57.

Kahl, Joseph A. 1953. "Educational and Occupational Aspirations of Common Man Boys." *Harvard Educational Review* 23:186–203.

———. 1957. *The American Class Structure*. 1st ed. New York: Rinehart.

Kahl, Joseph A. and James A. Davis. 1955. "A Comparison of Indexes of Socio-Economic Status." *American Sociological Review* 20:317–325.

Kane, Thomas. 2004. "College Going and Inequality." In *Social Inequality*, edited by K. Neckerman. New York: Russell Sage Foundation.

Kanter, Rosabeth. 1977. *Men and Women of the Corporation*. New York: Basic Books.

Karabel, Jerome and A. H. Halsey, eds. 1977. *Power and Ideology in Education*. New York: Oxford University Press.

Kassalow, Everett. 1978. "How Some European Nations Avoid U.S. Levels of Industrial Conflict." *Monthly Labor Review* 101(April):97.

Katz, Lawrence F. and Kevin M. Murphy. 1992. "Changes in Relative Wages, 1963–1987: Supply and Demand Factors." *Quarterly Journal of Economics* 107:35–78.

Kaus, Mickey. 1992. *The End of Equality*. New York: Basic Books.

Kaysen, Carl. 1957. "The Social Significance of the Modern Corporation." *American Economic Review* 47:311–319.

Keister, Lisa A. 2005. *Getting Rich: America's New Rich and How They Got That Way.* New York: Cambridge University Press.

Kelly, Kitty. 2004. *The Family: The Real Story of the Bush Family.* New York: Doubleday.

Kennickell, Arthur. 2006. "Currents and Undercurrents: Changes in the Distribution of Wealth, 1980–2004." *Federal Reserve Working Paper No. 2006-13.*

Kennickell, Arthur B., Douglas A. McManus, and R. Louise Woodburn. 1996. "Weighting Design for the 1992 Survey of Consumer Finances." Federal Reserve. Available at http://www.federalreserve.gov/Pubs/oss/oss2/papers/weight92.pdf.

Kerr, Clark and Abraham Siegel. 1954. "The Interindustry Propensity to Strike—An International Comparison." In *Industrial Conflict,* edited by A. Kornhauser et al. New York: McGraw-Hill.

Kichen, Steve et al. 1996. "The Private 500." *Forbes.* December 2.

Kingston, Paul W. 2000. *The Classless Society.* Palo Alto, CA: Stanford University Press.

Klinkner, Philip, ed. 1996. *Midterm: The Elections of 1994 in Context.* Boulder, CO: Westview.

Kodrzycki, Yolanda K. 1996. "Labor Market and Earnings Inequality: A Status Report." *New England Economic Review,* May/June 1996:11–25.

Koenig, Thomas. 1980. "Corporate Support for Political Contribution Disclosure." Unpublished paper presented at the American Sociological Association, New York.

Kohn, Melvin L. 1969. *Class and Conformity: A Study in Values.* Homewood, IL: Dorsey.

———. 1976. Social Class and Parental Values: Another Conformation of the Relationship. *American Sociological Review* 41:538–545.

———. 1977. *Class and Conformity.* 2nd ed. Chicago: University of Chicago Press.

Kohn, Melvin L. and Carmi Schooler. 1983. *Work and Personality: An Inquiry into the Impact of Social Stratification.* Norwood, NJ: Ablex.

Komarovsky, Mirra. 1946. "The Voluntary Associations of Urban Dwellers." *American Sociological Review* 11:689–698.

———. 1962. *Blue Collar Marriage.* New York: Vintage.

Kosman, Barry A. and Seymour P. Lachman. 1993. *One Nation Under God.* New York: Harmony Books.

Krugman, Paul and Robert Lawrence. 1994. "Trade, Jobs, and Wages." *Scientific American* April:44–49.

Lamont, Michele. 1992. *Money, Morals and Manners: The Culture of the French and the American Upper-Middle Class.* Chicago: University of Chicago Press.

———. 2000. *The Dignity of Working Men: Morality and the Boundaries of Race, Class, and Immigration.* Boston: Harvard University Press.

Landecker, Werner S. 1981. *Class Crystallization.* New Brunswick, NJ: Rutgers University Press.

Langerfeld, Steven. 1981. "To Break a Union." *Harpers* 262(May):16–21.

Langman, Lauren. 1987. "Social Stratification." In *Handbook of Marriage and the Family,* edited by M. Sussman and S. Steinmetz. New York: Plenum.

Lareau, Annette. 2003. *Unequal Childhoods: Class, Race, and Family Life.* Berkeley: University of California Press.

Lassiter, Luke Eric et al. 2004. *The Other Side of Middletown: Exploring Muncie's African American Community.* Walnut Creek, CA: Altamira Press.

Laumann, Edward O. 1966. *Prestige and Association in an Urban Community.* Indianapolis, IN: Bobbs-Merrill.

———. 1973. *Bonds of Pluralism: The Form and Substance of Urban Social Networks.* New York: Wiley.

Lavelle, Louis. 2001. "Special Report: Executive Pay." *BusinessWeek,* April 16.

Lazerow, Michael. 1995. "Millionaires Now 14% of House." *Roll Call*, July 10.

Leahy, Robert. 1981. "The Development of the Conception of Economic Inequality." *Child Development* 52:523–532.

———. 1983. *The Child's Construction of Social Inequality.* New York: Academic.

Lemann, Nicholas. 2000. *The Big Test: The Secret History of the American Meritocracy.* New York: Farrar, Straus & Giroux.

LeMasters, E. E. 1975. *Blue-Collar Aristocrats: Life-Styles at a Working-Class Tavern.* Madison: University of Wisconsin Press.

Lenski, Gerhard. 1954. "Status Crystallization: A Non-Vertical Dimension of Social Status." *American Sociological Review* 19:405–413.

———. 1966. *Power and Privilege: A Theory of Social Stratification.* New York: McGraw-Hill.

Lerner, Robert, Althea Nagai, and Stanley Rothman. 1996. *American Elites.* New Haven, CT: Yale University Press.

Levine, Donald M. and Mary Jo Bane, eds. 1975. *The "Inequality" Controversy: Schooling and Distributive Justice.* New York: Basic Books.

Levitan, Sar A. 1990. *Programs in Aid of the Poor.* 6th ed. Baltimore: Johns Hopkins University Press.

Levitan, Sar and Isaac Shapiro. 1987. *Working but Poor: America's Contradiction.* Baltimore: Johns Hopkins University Press.

Levitan, Sar A. and Robert Taggart. 1976. *The Promise of Greatness.* Cambridge, MA: Harvard University Press.

Levy, Frank. 1995. "Incomes and Income Inequality." In *State of the Union: America in the 1990s*, Vol. 1, edited by R. Farley. New York: Russell Sage Foundation.

Levy, Frank and Richard J. Murnane. 1992. "U.S. Earnings Levels and Earnings Inequality: A Review of Recent Trends and Proposed Explanations." *Journal of Economic Literature,* 30:1333–1381.

Lewis, Neil. 1996. "This Mr. Smith Gets His Way in Washington." *New York Times,* Oct. 12.

Lewis, Sinclair. 1922. *Babbitt.* New York: Harcourt Brace Jovanovich.

Link, Arthur S. and William Cotton. 1973. *American Epoch.* Vol. 1. 4th ed. New York: Knopf.

Lipset, Seymour Martin. 1960. *Political Man.* New York: Doubleday.

———. 1981. *Political Man.* Expanded ed. Baltimore: Johns Hopkins University Press.

Litwack, Leon, ed. 1962. *The American Labor Movement.* Englewood Cliffs, NJ: Prentice Hall.

Lopata, Helena Z. et al. 1980. "Spouses' Contributions to Each Other's Roles." In *Dual-Career Couples,* edited by F. Pepitone-Rockwell. Beverly Hills, CA: Sage.

Lord, Walter. 1955. *A Night to Remember.* New York: Henry Holt.

Lorwin, Lewis L. 1933. *The American Federation of Labor.* Washington, DC: Brookings Institution.

LTV Corporation. 1990. A Guide to the 102nd Congress: 1st Session Datebook/Calendar. Washington, DC: Author.

Lubell, Samuel. 1956. *The Future of American Politics.* 2nd ed. Garden City, NY: Doubleday Anchor.

Lucas, Samuel Roundfield. 1999. *Tracking Inequality: Stratification and Mobility in American High Schools.* New York: Teachers College Press.

Lundberg, Ferdinand. 1968. *The Rich and the Super-Rich.* New York: Lyle Stuart.

Lynd, Robert S. and Helen Merrell Lynd. 1929. *Middletown.* New York: Harcourt Brace Jovanovich.

———. 1937. *Middletown in Transition.* New York: Harcourt Brace Jovanovich.

Mackenzie, Gavin. 1973. *The Aristocracy of Labor: The Position of Skilled Craftsmen in the American Class Structure.* New York: Cambridge University Press.

Madison, James (with Alexander Hamilton and John Jay). [1787] 1961. *The Federalist Papers*. New York: New American Library.

Makinson, Larry. 1990. *Open Secrets: The Dollar Power of PACs in Congress*. Washington, DC: Congressional Quarterly Press.

Makinson, Larry and Joshua Goldstein. 1996. *Open Secrets: The Encyclopedia of Congressional Money and Politics*. 4th ed. Washington, DC: Congressional Quarterly Press.

Malbin, Michael J., ed. 2006. *The Election After Reform: Money, Politics, and the Bipartisan Campaign Reform Act*. Lanham, MD: Rowman & Littlefield.

Manza, Jeff and Clem Brooks. 1999. *Social Cleavages and Political Change: Voter Alignments and U.S. Party Coalitions*. New York: Oxford University Press.

Marcus, Ruth and Charles Babcock. 1997. "The System Cracks Under the Weight of Cash." *The Washington Post*. February 9.

Marx, Karl. 1979. *The Marx-Engels Reader*, edited by Robert C. Tucker. 2nd ed. New York: Norton.

Massey, Douglas. 1996. "The Age of Extremes: Concentrated Affluence and Poverty in the Twenty-First Century." *Demography* 33:395–412.

Massey, Douglas and Mary J. Fisher. 2003. "The Geography of Inequality in the United States, 1950–2000." In *Brookings-Wharton Papers on Urban Affairs 2003*, edited by W. G. Gale and J. T. Pack. Washington, DC: Brookings Institution.

Mazumder, Bhashkar. 2005. "The Apple Falls Even Closer to the Tree Than We Thought." In *Unequal Chances: Family Background and Economic Success*, edited by S. Bowles, H. Gintis, and M. Groves. Princeton, NJ: Princeton University Press.

McCall, Leslie. 2001. *Complex Inequality: Gender, Class, and Race in the New Economy*. New York: Routledge.

McCarty, Nolan, Keith T. Poole, and Howard Rosenthal. 2006. *Polarized America: The Dance of Ideology and Unequal Riches*. Cambridge, MA: MIT Press.

McLeod, Jay. 2004. *Ain't No Makin' It: Leveled Aspirations in a Low-Income Neighborhood*. Revised ed. Boulder, CO: Westview.

Miller, Herman P. 1971. *Rich Man, Poor Man*. Rev. ed. New York: Thomas Y. Crowell.

Mills, C. Wright. 1951. *White Collar*. New York: Oxford University Press.

———. 1956. *The Power Elite*. New York: Oxford University Press.

———. 1968. "Comment on Criticism." In *C. Wright Mills and the Power Elite*, edited by G. W. Domhoff and H. B. Ballard. Boston: Beacon.

Mintz, Beth. 1975. "The President's Cabinet, 1897–1972." *Insurgent Sociologist* 5(3):131–149.

Mishel, Lawrence, Jared Bernstein, and Sylvia Allegretto. 2007. *The State of Working America, 2006/2007*. Ithaca, NY: Cornell University Press.

Mishel, Lawrence, Jared Bernstein, and John Schmitt. 2001. *The State of Working America, 2000/2001*. Ithaca, NY: Cornell University Press.

Mortenson, Thomas G. 1991. *Equity of Higher Educational Opportunity for Women, Black, Hispanic, and Low Income Students*. Iowa City, IA: American College Testing Program.

———. 2001. "Family Income and Higher Education Opportunity, 1970–2000." *Postsecondary Education Opportunity*, October.

Murray, Charles A. 1984. *Losing Ground: American Social Policy*. New York: Basic Books.

Nagle, John. 1977. *System and Succession: The Social Bases of Political Elite Recruitment*. Austin: University of Texas Press.

Naimark, Hedwin. 1981. *The Development of the Understanding of Social Class*. Doctoral dissertation, New York University.

Nakao, Keiko and Judith Treas. 1990. "Revised Prestige Scores for All Occupations." Chicago: National Opinion Research Center (unpublished paper).

National Commission on Children. 1991. *Beyond Rhetoric: A New American Agenda for Children and Families*. Washington, DC: U.S. Government Printing Office.

National Opinion Research Center (NORC). 1953. "Jobs and Occupations: A Popular Evaluation." In *Class, Status, and Power,* edited by R. Bendix and S. M. Lipset. Glencoe, IL: Free Press.

Neckerman, Kathryn M., ed. 2004. *Social Inequality.* New York: Russell Sage Foundation.

Newman, Katherine S. 1988. *Falling from Grace: The Experience of Downward Mobility in the American Middle Class.* New York: Free Press.

New York Times. 2001. "The New York Times Almanac 2002." New York: Penguin.

New York Times (Eric Dash). 2007. "Executive Pay: A Special Report." April 8.

Nord, Mark, Margaret Andrews, and Steven Carlson. 2003. *Household Food Security in the United States, 2002.* U.S. Department of Agriculture. Food Assistance and Nutrition Research Report No. 35.

Oakes, Jeannie. 1985. *Keeping Track: How High Schools Structure Inequality.* New Haven, CT: Yale University Press.

Oliver, Melvin and Thomas Shapiro. 1995. *Black Wealth/White Wealth: A New Perspective on Racial Inequality.* New York: Routledge.

O'Neill, June. 2006. Statement. *Hearings on 1996 Welfare Overhaul. U.S. House of Representatives.* Committee on Ways and Means. July 19.

Ornati, Oscar. 1966. *Poverty Amidst Affluence.* New York: Twentieth Century Fund.

Orshansky, Mollie. 1974. "How Poverty Is Measured." In *Sociology of American Poverty,* edited by J. Huber et al. Cambridge, MA: Schenkman.

Ossowski, Stanislaw. 1963. *Class Structure in the Social Consciousness.* New York: Free Press.

Osterman, Paul. 1991. "Gains From Growth? The Impact of Full Employment on Poverty in Boston." In *The Urban Underclass,* edited by C. Jencks and P. Peterson. Washington, DC: Brookings Institution.

Ostrander, Susan. 1984. *Women of the Upper Class.* Philadelphia: Temple University Press.

Pakulski, Jan and Malcolm Waters. 1996. *The Death of Class.* Thousand Oaks, CA: Sage.

Parrott, Sharon. 2006. Statement. *Hearings on 1996 Welfare Overhaul. U.S. House of Representatives.* Committee on Ways and Means. July 19.

Pen, Jan. 1971. *Income Distribution.* London: Allen Lane.

Persell, Caroline Hodges. 1977. *Education and Inequality.* New York: Free Press.

Phillips, Katherin. 2001. "The Earned Income Tax Credit: Knowledge Is Money." *Political Science Quarterly* 116:413–424.

Phillips, Kevin. 1990. *The Politics of Rich and Poor: Wealth and the American Electorate in the Reagan Aftermath.* New York: Random House.

———. 2002. *Wealth and Democracy: A Political History of the American Rich.* New York: Broadway Books.

Pianin, Eric. 1997. "How Business Found Benefits in Wage Bill." *The Washington Post,* February 11.

Piven, Frances Fox and Richard A. Cloward. 1971. *Regulating the Poor: The Functions of Public Welfare.* New York: Pantheon.

Polsby, Nelson. 1970. "How to Study Community Power: The Pluralist Alternative." In *The Structure of Community Power,* edited by M. Aiken and P. E. Mott. New York: Random House.

Pomper, Gerald M. 1989. "The Presidential Power." Pp. 153–176 in *The Election of 1988,* edited by G. M. Pomper. Chatham, NJ: Chatham House.

Rainwater, Lee. 1965. *Family Design: Marital Sexuality, Family Size, and Contraception.* Chicago: Aldine.

———. 1974. *What Money Buys.* New York: Basic Books.

Ramsey, Patricia. 1991. "Young Children's Awareness and Understanding of Social Class Differences." *Journal of Genetic Psychology* 152:71–82.

Reischauer, Robert. 1997. *Setting National Priorities: Budget Choices for the Next Century.* Washington, DC: Brookings Institution.

Reiss, Albert. 1961. *Occupations and Social Status.* Glencoe, IL: Free Press.

Reissman, Leonard. 1954. "Class, Leisure and Participation." *American Sociological Review* 19:74–84.

Rieder, Jonathan. 1985. *Canarsie: The Jews and Italians of Brooklyn Against Liberalism.* Cambridge, MA: Harvard University Press.

Riesman, David. 1953. *The Lonely Crowd: A Study of the Changing American Character.* New Haven, CT: Yale University Press.

Ritter, Kathleen, and Lowell Hargens. 1975. "Occupational Positions and Class Identifications of Married Women." *American Journal of Sociology* 80:934–948.

Ritzer, George. 1996. *The McDonaldization of Society: An Investigation Into the Changing Character of Contemporary Social Life.* Thousand Oaks, CA: Pine Forge.

Rose, Stephen J. 1993. "Declining Family Incomes in the 1980s: New Evidence From Longitudinal Data." *Challenge* November/December:29–36.

———. 1992. *Social Stratification in the United States: The American Profile Poster Revised and Expanded.* New York: New Press.

———. 1994. *On Shaky Ground: Rising Fears About Income and Earnings.* Washington, DC: National Commission for Employment Policy.

———. 2007. *Social Stratification in the United States: The American Profile Poster.* Revised and updated ed. New York: New Press.

Rosen, Ellen Israel. 1987. *Bitter Choices: Blue-Collar Women In and Out of Work.* Chicago: University of Chicago Press.

Ross, Howard. 1968. "Economic Growth and Change in the United States under Laissez-Faire: 1870–1929." In *The Age of Industrialization in America,* edited by F. C. Jaher. New York: Free Press.

Rossi, Peter H. 1989. *Down and Out in America: The Origins of Homelessness.* Chicago: University of Chicago Press.

Rubin, Lillian Breslow. 1976. *Worlds of Pain: Life in the Working-Class Family.* New York: Basic Books.

———. 1994. *Families on the Faultline: America's Working Class Speaks About the Family, the Economy, Race, and Ethnicity.* New York: HarperCollins.

Rubin, Z. 1968. "Do American Women Marry Up?" *American Sociological Review* 5:750–760.

Ryscavage, Paul et al. 1992. "The Impact of Demographic, Social, and Economic Change on the Distribution of Income." In U.S. Census Bureau, *Studies in the Distribution of Income.* Washington, DC: U.S. Government Printing Office.

Sachs, Jeffrey and Howard Shatz. 1994. "Trade and Jobs in U.S. Manufacturing." *Brookings Papers on Economic Activity* Number 1.

Sawhill, Isabel. 1989. "The Underclass: An Overview." *The Public Interest* 96(Summer):3–15.

Schiller, Bradley. 1989. *The Economics of Poverty and Discrimination.* 5th ed. Englewood Cliffs, NJ: Prentice Hall.

Schlesinger, Arthur J., Jr. (1945). *The Age of Jackson.* Boston: Little, Brown.

Schreiber, E. M. and G. T. Nygreen. 1970. "Subjective Social Class in America: 1945–1968." *Social Forces* 45:348–356.

Schwartz, Michael, ed. 1987. *The Structure of Power in America: The Corporate Elite as a Ruling Class.* New York: Holmes & Meier.

Sernau, Scott and Jonnie Griffin, eds. 2004. *Social Stratification Courses: Syllabi and Instructional Materials.* 3rd ed. Washington, DC: American Sociological Association.

———. 2000. *Social Stratification Courses: Syllabi and Instructional Materials.* 4th ed. Washington, DC: American Sociological Association.

Sewell, William H. and Robert M. Hauser. 1975. *Education, Occupation, and Earnings.* New York: Academic.

Sewell, William H. and Vimal P. Shah. 1977. "Socioeconomic Status, Intelligence, and the Attainment of Higher Education." In *Power and Ideology in Education,* edited by J. Karabel and A. H. Halsey. New York: Oxford University Press.

Shapiro, Thomas. 2004. *The Hidden Cost of Being African American: How Wealth Perpetuates Inequality.* New York: Oxford University Press.

Shipler, David. 2005. *The Working Poor: Invisible in America.* New York: Random House.

Shostak, Arthur and William Gomberg, eds. 1964. *The Blue Collar World: Studies of the American Worker.* Englewood Cliffs, NJ : Prentice Hall.

Simkus, Albert. 1978. "Residential Segregation by Occupation and Race." *American Sociological Review* 43:81–93.

Simmons, Robert G. and Morris Rosenberg. 1971. "Functions of Children's Perceptions of the Stratification System." *American Sociological Review* 36:235–249.

Smith, David H. and Jacqueline Macaulay. 1980. *Participation in Social and Political Activities.* San Francisco: Jossey-Bass.

Smith, James P. and Finis R. Welch. 1989. "Black Economic Progress After Myrdal." *Journal of Economic Literature* 27:519–564.

Smith, Tom W. 1987. "That Which We Call Welfare by Any Other Name Would Smell Sweeter: An Analysis of the Impact of Question Wording on Response Patterns." *Public Opinion Quarterly* 51:75–83.

Snow, David and Leon Anderson. 1993. *Down on Their Luck: Homeless Street People.* Berkeley: University of California Press.

Sorauf, Frank. 1992. *Inside Campaign Finance: Myths and Realities.* New Haven, CT: Yale University Press.

Sorensen, Annemette. 1994. "Women, Family, and Class." *Annual Review of Sociology* 20:27–47.

Stack, Carol B. 1974. *All Our Kin.* New York: Harper & Row.

Statesman's Yearbook. 1979–1990. New York: St. Martin's.

Stendler, Celia Burns. 1949. *Children of Brasstown: Their Awareness of the Symbols of Social Class.* Urbana: University of Illinois Press.

Stevens, Mitchell. 2007. *Creating a Class: College Admissions and the Education of Elites.* Cambridge, MA: Harvard University Press.

Sussman, Marvin and Suzanne Steinmetz. 1987. *Handbook of Marriage and the Family.* New York: Plenum.

Sweezy, Paul. 1968. "Power Elite or Ruling Class?" In *C. Wright Mills and the Power Elite,* edited by G. W. Domhoff and H. B. Ballard. Boston: Beacon.

Szymanski, Albert. 1978. *The Capitalist State and the Politics of Class.* Cambridge, MA: Winthrop.

Teixeira, Ruy. 1992. *The Disappearing American Voter.* Washington, DC: Brookings Institution.

Teixeira, Ruy and Joel Rogers. 2000. *America's Forgotten Majority: Why the White Working Class Still Matters.* New York: Basic Books.

Terkel, Studs. 1974. *Working: People Talk About What They Do All Day and How They Feel About It.* New York: Avon.

Thompson, E. P. 1963. *The Making of the English Working Class.* New York: Vintage.

Treas, Judith and Ramon Torrecilha. 1995. "The Older Population." In *The State of the Union: America in the 1990s; Volume 2: Social Trends.* New York: Russell Sage Foundation.

Treiman, Donald. 1977. *Occupational Prestige in Comparative Perspective.* New York: Academic.

Treiman, Donald, and Patricia Roos. 1983. "Sex and Earnings in Industrial Society: A Nine-Nation Comparison." *American Journal of Sociology* 89:612–650.

Tuchman, Gaye, ed. 1974. *The TV Establishment: Programming for Power and Profit.* Englewood Cliffs, NJ: Prentice Hall.

Tutor, Jeannette. 1991. "The Development of Class Awareness in Children." *Social Forces* 49:470–476.

United for a Fair Economy, ed. 2004. *The Wealth Inequality Reader.* Cambridge, MA: Dollars & Sense, Economic Affairs Bureau.

U.S. Census. 2003. *Net Worth and Asset Ownership of Households 1998 and 2000.* Washington, DC: U.S. Census Bureau.

U.S. Census Bureau. 1973–2001. *Statistical Abstract of the United States.* Various editions.

———. 1975. *Historical Statistics of the United States.* Bicentennial ed. 2 Parts.

———. 1976. *Bicentennial Statistics, Pocket Data Book, USA.*

———. 1980a. "The Social and Economic Status of the Black Population in the United States: An Historical View, 1790–1978." *Current Population Reports,* Special Studies Series P-23, no. 80.

———. 1980b. *Statistical Abstract of the United States: 1980.*

———. 1983. *Statistical Abstract of the United States: 1984.*

———. 1984a. "Money Income and Poverty Status of Families and Persons in the United States: 1983—Advance Data." *Current Population Reports.* Series P-60, no. 145.

———. 1984b. *1980 Census of Population,* United States Summary, Section A: United States. Characteristics of the Population. Series PC80-1-D1-A.

———. 1990a. *Trends in Income, by Selected Characteristics: 1947 to 1988.* Current Population Reports. Series P-60, no. 167.

———. 1990b. *Transition in Income and Poverty Status: 1985–86.* Series P-70, no. 18.

———. 1991. *Transition in Income and Poverty Status: 1987–88.* Series P-70, no. 24.

———. 1992. *Workers with Low Earnings: 1964–1990.*

———. 1993. *Money Income of Households, Families, and Persons in the United States: 1992.* Series P60-184.

———. 1994. *Dynamics of Economic Well-Being: Labor Force and Income, 1990 to 1992.* Series P70-40.

———. 1995. *Child Support for Custodial Mothers and Fathers: 1991.* Series P60-187.

———. 1996a. *Income, Poverty, and Valuation of Noncash Benefits: 1994.* Series P60-189.

———. 1996b. "A Brief Look at Postwar U.S. Income Inequality." *Current Population Reports.* Series P60-191.

———. 1999. *Experimental Poverty Measures: 1990–1997.*

———. 1999a. *Experimental Poverty Measures: 1998.*

———. 2000. *Poverty Among Working Families: Findings From Experimental Poverty Measures, 1998.*

———. 2001a. *Experimental Poverty Measures: 1999.*

———. 2001b. *Home Computers and Internet Use in the United States: August 2000.*

———. 2001c. *Household Net Worth and Asset Ownership: 1995.*

———. 2001d. *Poverty in the United States: 2000.*

———. 2006. *Custodial Mothers and Fathers and Their Child Support: 2003.*

U.S. Department of Housing and Urban Development. 1999. *Homelessness: Programs and the People They Serve.*

U.S. Department of Housing and Urban Development. 2007. *The Annual Homeless Assessment Report to Congress.*

U.S. Department of Justice. 2001. *Intimate Partner Violence and Age of Victim, 1993–99.*

U.S. Department of Labor. 1994. *Report on the American Workforce.*

———. 2001a. *Consumer Expenditures in 1999.* May.

———. 2001b. *Employment and Earnings.* January.

U.S. House of Representatives, Committee on Ways and Means. 1991a. *Overview of the Federal Tax System.* WMCP-102-7.

———. 1991b. *Background Material and Data on Programs Within the Jurisdiction of the Committee on Ways and Means.*

———. 1992. *Overview of Entitlement Programs: Background Material and Data on Programs Within the Jurisdiction of the Committee on Ways and Means.* May 15.

———. 2000. *Background Material and Data on Programs Within the Jurisdiction of the Committee on Ways and Means.*

U.S. Internal Revenue Service. 1988. *Statistics of Income.* Fall.

———. 1995. *Statistics of Income.* Spring.

———. 2000. *Statistics of Income.* Spring.

———. 2007. *Statistics of Income.* Winter.

Useem, Michael 1996. *Investor Capitalism: How Money Managers Are Changing the Face of Corporate America.* New York: Basic Books.

Vanfossen, Beth Ensminger. 1977. "Sexual Stratification and Sex-Role Socialization." *Journal of Marriage and the Family* 39:563–574.

Veblen, Thorstein. [1899] 1934. *The Theory of the Leisure Class.* New York: Modern Library.

Venkatesh, Sudhir Alladi. 2000. *American Project: The Rise and Fall of a Modern Ghetto.* Cambridge, MA: Harvard University Press.

Verba, Sidney, Kay Lehman Schlozman, and Henry E. Brady. 2004. "Political Equality: What Do We Know About It?" In *Social Inequality,* edited by K. M. Neckerman. New York: Russell Sage Foundation.

Warner, W. Lloyd and Paul S. Lunt. 1941. *The Social Life of a Modern Community.* New Haven, CT: Yale University Press.

Warner, W. Lloyd et al. 1949a. *Democracy in Jonesville.* New York: Harper & Row.

———. 1949b. *Social Class in America.* Chicago: Science Research Associates.

———. 1973. *Yankee City.* Abridged ed. New Haven, CT: Yale University Press.

Wattenberg, Ben J. 1974. *The Real America.* Garden City, NY: Doubleday.

Weber, Max. 1946. *From Max Weber: Essays in Sociology,* edited by H. H. Gerth and C. W. Mills. New York: Oxford University Press.

Weissman, Stephen and Ruth Hassan 2006. "527 Groups and BCRA." In *The Election After Reform: Money, Politics, and the Bipartisan Campaign Reform Act,* edited by Michael J. Malbin. Lanham, MD: Rowman & Littlefield.

Wetzel, James R. 1995. "Labor Force, Unemployment, and Earnings." In *State of the Union: America in the 1990s, Volume 1: Economic Trends,* edited by R. Farley. New York: Russell Sage Foundation.

Who's Who in America. 1980. 41st ed. Chicago: Marquis.

Who's Who in American Politics. 1979. 7th ed. New York: Bowker.

Whyte, Martin King. 1990. *Dating, Mating, and Marriage.* New York: Aldine de Gruyter.

Whyte, William H. 1952. *Is Anybody Listening?* New York: Simon & Schuster.

Wilson, William J. 1980. *The Declining Significance of Race.* 2nd ed. Chicago: University of Chicago Press.

———. 1987. *The Truly Disadvantaged: The Inner City, the Underclass, and Public Policy.* Chicago: University of Chicago Press.

———. 1991. "Public Policy Research and the Truly Disadvantaged." In *The Urban Underclass,* edited by C. Jencks and P. Peterson. Washington, DC: Brookings Institution.

———. 1996. *When Work Disappears: The World of the New Urban Poor.* New York: Knopf.

Wilson, William Julius and Kathryn Neckerman. 1986. "Poverty and Family Structure: The Widening Gap Between Evidence and Public Policy Issues." In *Fighting Poverty: What*

Works and What Doesn't, edited by S. Danziger and D. H. Weinberg. Cambridge, MA: Harvard University Press.

Wolfe, Tom. 1987. *The Bonfire of the Vanities.* New York: Farrar, Straus & Giroux.

Wolff, Edward. 1993. "The Structure of Wealth Inequality: A Report to the Twentieth Century Fund." Unpublished paper.

———. 1996. "Trends in Household Wealth, 1983–1992." Report Submitted to the Department of Labor. Unpublished paper.

———. 1998. "Recent Trends in the Size Distribution of Household Wealth." *Journal of Economic Perspectives* 12:131–150.

Wolfinger, Raymond E. 1973. *The Politics of Progress.* Englewood Cliffs, NJ: Prentice Hall.

Wool, Harold. 1976. *The Labor Supply for Lower-Level Occupations.* New York: Praeger.

Wright, James. 1989. *Address Unknown: The Homeless in America.* New York: Aldine de Gruyter.

Zeitlin, Maurice. 1980. *Classes, Class Conflict and the State.* Cambridge, MA: Winthrop.

Zweigenhaft, Richard L. and G. William Domhoff. 1998. *Diversity in the Power Elite: Have Women and Minorities Reached the Top?* New Haven, CT: Yale University Press.

Note on Statistical Sources

Most of the statistics cited in the text and analyzed in tables and graphs come from the U.S. Census Bureau or the Bureau of Labor Statistics. In recent years, both agencies have placed mountains of information on the Web, while apparently reducing their publication programs. To avoid cluttering the text with endless Web site and publication references, I have eliminated most citations for statistics derived from the Census Bureau's "March Current Population Survey" (the main source of income, poverty, and occupation data) and other standard series administered by these agencies. I have done the same for other statistical series that appear regularly in the Census Bureau's *Statistical Abstract of the United States.* However, information drawn from special publications of government agencies, rather than their regular series, are cited in the text and listed in the bibliography. Where necessary, government Web sites are cited by the agency's name (Census Bureau, Justice Dept., and so forth). All were accessed in 2006 and 2007. The Census Bureau site (www.census.gov) has links to other relevant sites, including the important Bureau of Labor Statistics site.

Credits

Permission to reprint excerpted material has been granted for the following:

Index

Pages with t indicate a table, f indicate a figure, and n indicate a note.

Committee for Industrial
    Organization (CIO), 195
Communism, 4
Communist Manifesto (Marx), 5
Conformity, 97–98
Congress, U.S., 166–167
Congress of Industrial
    Organizations (CIO), 195
Constraint, 102
Cook, Philip, 63–64
Corporations, 158–161, 165–167,
    171–175. *See also* Power
Correlation, simple, 133, 243
Council of Economic Advisors, 208
Cultivated growth of children, 99–102
Cultural capital, 94. *See also* Capital
Cutting consistency, 243

Dahl, Robert, 152
Davis, Allison, 24–26
Deep South (Davis, Gardner,
    and Gardner), 24–26
Democratic Party, 188–191
Depression, Great, 196, 205
Deprivation, 181
Direct relief, 205
Distribution of income, 69–73, 75–79,
    87–88, 237, 243
Distribution of wealth, 83–86,
    237, 243
Dividends, 243
Divorce, 221–222
Domestic violence, 111–113
Domhoff, G. William, 157
Downsizing, 62, 243
Dual careers, 106
Duncan, Otis Dudley, 134–137
Dye, Thomas, 155

Earned income tax credit (EITC),
    79, 218, 223–224, 243
Earnings
    defined, 65, 243
    during Age of Growing Inequality,
        57–59
    CEOs, 57–58, 63, 161
    corporate strategies to reduce labor
        costs and, 61–62
    equity and, 144–145

gap, 57, 59t, 60–61, 144–145,
    236–237, 238–239
gender and, 54, 57–58, 144–145
poverty and, 15, 16f, 219–220
race and, 55
by women, 81–82
See also Income
Economic capital, 94. *See also* Capital
Economic insecurity, 181
Economic restructuring, 60
Economics. *See* Earnings; Income; Power;
    specific classes; Wealth
Education
    association patterns and, 102–103
    income and, 138–139, 140–141
    parental values and, 98–102
    prep schools for, 162
    social mobility and, 138–139,
        140–145, 164
    socioeconomic status, effect on,
        135–137
    upper–middle class, importance to, 233
Edwards schema for occupations, 47–49
Effective tax rates, 79–80, 90, 243
EITC (Earned income tax credit). *See*
    Earned income tax credit (EITC)
Elderly, 81
Elections. *See* Politics
Elite
    cohesion, 152–154, 155, 163, 244
    defined, 149, 244
    perspective on power, 150–154,
        157–158
Employment. *See* Occupation;
    Unemployment
End of Equality, The (Kaus), 117
Endogamy, 103–104, 162, 244
Entitlement, 101–102, 205–206,
    217, 244
Equality of opportunity, 133
Establishment, the, 164–165, 244
Estate tax, 90–91, 167, 168, 172
Ethnicity, 44. *See also* Minorities; Race
Executive branch of government,
    150–151, 165–166

Family
    class determination and, 11–12
    defined, 11n, 12, 244